Study Guide For
Chemical
Principles
Fifth Edition

Masterson, Slowinski and Stanitski

Raymond L. Boyington

Professor of Chemistry
University of Connecticut
Storrs, Connecticut

SAUNDERS GOLDEN SUNBURST SERIES

SAUNDERS COLLEGE PUBLISHING 1981
Philadelphia

SAUNDERS COLLEGE PUBLISHING/CBS
Educational and Professional
Publishing, A Division of CBS, Inc.
Address Orders to: 383 Madison Ave.,
New York, NY 10017
Address Editorial correspondence to:
West Washington Square
Philadelphia, PA 19105

STUDY GUIDE TO ACCOMPANY CHEMICAL PRINCIPLES   ISBN O O3 O582767 8

1234 OO5 987654321

CBS COLLEGE PUBLISHING

Preface

This study guide has been written as an aid to the student of general chemistry — to supplement a textbook or lecture series, or to guide independent study. The chapter sequence and selection of topics parallel that of *Chemical Principles* (fifth edition) by Masterton, Slowinski, and Stanitski, in both standard and SI versions.

TO THE STUDENT

Each chapter in this guide begins with a set of *Questions to Guide Your Study*. These are meant to suggest some of the things you should be looking for in studying the topics of the chapter and the corresponding text and lecture material. You should add your own questions to this partial listing. Next follows a list of concepts and mathematical operations *You Will Need to Know* in order to get the most out of your study. The concepts and math are usually topics carried over from previous chapters which may need reviewing before you go on to the topics of the current chapter.

Basic Skills lists the new concepts and problem-solving techniques that must be mastered before you can work the problems in the text or on an examination. Each skill is illustrated by one or more worked examples, either in this guide or the text, or both. Illustrative, end-of-chapter problems in *Chemical Principles* have been similarly cataloged in the text itself.

The skills and factual content of chemistry are then tested in the next section of the guide chapter. Each *Self-Test* consists of 20 objective questions (true/false and multiple choice, each with only one best answer) and five problems. The entire test should require about one hour to work. The test may be considered as a typical examination for the given chapter material. In fact, each test is similar to the examinations given at the University of Connecticut over the last ten years. Of course, no one test can possibly cover all the topics considered in the text chapter or in a lecture. Nor can you expect the level of difficulty to be exactly the same from one test to every other. Still, working the tests in this guide should be very useful practice in taking exams as well as in helping you evaluate your progress.

Work the Self-Test under exam-like conditions. Time yourself. Do not waste time on a particular question or problem. If you are stuck, go on to the next and come back later if there is time. On a multiple choice question, be sure to read through all the answers before making a choice. Spend no more than five minutes setting up a problem. Check all units for consistency. See that you have indeed solved for what the problem asked. Use no other aids than a Periodic Table and an electronic calculator.

Grade your test, using the *Self-Test Answers* that immediately follow the test. (Don't peek at them while working the test!) For the objective questions, 14 correct answers (70%) should be about "average." For the problems, if the number correct is

0-1	Get help!
2	You are making progress, but need to do more work.
3	About average. You might score a C on another exam.
4	Superior work: maybe worth a B+ or A-.
5	Help someone else!

Finally, following the answers to the test is a list of recommended readings. These *Selected Readings* have been chosen to provide more basic background material and other interesting and important applications, as well as more advanced study. Still other study aids have been listed at the end of this preface.

If you are enrolled in a course of lectures, perhaps with discussion and laboratory classes, consider the following suggestions for getting more out of your study. Try at least one you have not tried before.

– Prepare for the lectures! Skim through the text material you expect to be discussed. Read the section headings and introductory paragraphs; look at the tables and graphs; read the chapter summary. Try to get a general idea of what it is about. Read and add you own notes to the first two sections of the study guide chapter. After each lecture, go back and *work* through the text chapter. Solve for yourself the worked-out examples. Line by line, you should ask what is given, what is called for, what comes next, where you find the required data . . . . Don't simply "follow" the given solution! In this and all parts of your study, you must actively participate. Try to anticipate answers and even questions. Jot down anything that requires further explanation. Write out conclusions and summaries in your own words.

– Attend the lectures and take notes. The more thoroughly you have prepared for lectures, the fewer notes you will need to take and the more you can think about what is being said. It is much easier to understand something in a lecture than hours or days later when you read over your notes. And the more questions you can raise, even if only to yourself. And the more time you may have for other things.

Do not merely take notes – think about what is being said, step by step. Notes are a place for raising questions as well as recording answers to such questions as these: What is the lecturer's emphasis? What examples and applications are given? What are the other interpretations? What experiments can be done to support a particular point?

— Prepare for discussions (as well as labs!). Work through as many problems in the text as you have time for. Make note of anything requiring explanation. Think about your answers: Do they seem reasonable in light of what you already know? Do they suggest other problems? Get as much as possible out of discussions by asking your questions. Make your problems known as clearly and as early as possible. You usually do have a real influence on how beneficial a discussion class may be. Make it work for you by being prepared.

— Review regularly and frequently. This is where your notes and this guide should be most helpful. Look for connections between topics. Summarize sections and chapters for yourself. Tabulate formulas along with the conditions under which they may be used.

— Before taking exams, recall the special emphases made in lectures and discussions, as well as here and in the text. Try to construct questions like those raised in lectures and in the text. Be sure you are familiar with all the basic skills. Work some more problems. Work the test in this guide. Try to work tests that have been given before in the course.

— Read more chemistry, whether out of curiosity or from a feeling of helplessness. The more you do, the more rewarding it will be, and the less help you will need.

— Try to teach what you have learned. You will be surprised how it helps you organize what you know, as well as discover what you need to work on some more.

ADDITIONAL STUDY AIDS
Math Preparation and Problem-Solving Manuals, and General Study Aids

Apps, J. W., *Study Skills for Those Adults Returning to School*, New York, McGraw-Hill, 1978.

Barrow, G. M., *Understanding Chemistry*, New York, W. A. Benjamin, 1969.

Butler, I. S., *Relevant Problems for Chemical Principles*, Menlo Park, Calif., W. A. Benjamin, 1970.

Drago, R. S., *General Chemistry Problem Solving*, Voorhees, N.J., Spring Dale, 1979.

Edman, D. D., "Computers and Chemistry," *Chemistry* (January 1972), pp. 6–9.

Gibson, G. W., *Mastering Chemistry: A Problem Solving Guide for Introductory Chemistry*, Philadelphia, W. B. Saunders, 1975.

Johnson, K. J., *General Chemistry Examination Questions*, Minneapolis, Burgess, 1978.

Mallow, J. V., "Science Anxiety," *Chemistry* (October 1978), pp. 6–9.

Masterton, W. L., *Elementary Mathematical Preparation for General Chemistry,* Philadelphia, W. B. Saunders, 1974.

Musgrave, T. R., *Understanding Problems for Chemical Principles,* Philadelphia, W. B. Saunders, 1978.

O'Connor, R., *Solving Problems in Chemistry,* New York, Harper and Row, 1977.

Peters, E. I., *Problem Solving for Chemistry,* Philadelphia, W. B. Saunders, 1976.

Polya, G., *How to Solve It,* New York, Doubleday Anchor, 1957.

Sienko, M. J., *Chemistry Problems,* Menlo Park, Calif., W. A. Benjamin, 1972.

Smith, R. N., *Solving General Chemistry Problems,* San Francisco, W. H. Freeman, 1980.

Soltzberg, L., *BASIC and Chemistry,* Boston, Houghton Mifflin, 1975.

Tobias, S., *Overcoming Math Anxiety,* Boston, Houghton Mifflin, 1978.

Journals and Magazines

Chemical and Engineering News

Journal of Chemical Education

Scientific American

SciQuest (formerly called *Chemistry*)

Reprints of articles are often available free from the author; sometimes, for a small fee, from the publisher.

Audio-Visual Aids

Tapes, films, slides, programmed instruction materials, even lecture notes and other resources that might be used for self-instruction are often available but not used. Find out where they are and how to use them. Often they are available through a library or a departmental office.

Also consider:

O'Connor, R., *Topics-Aids: A Guide to Instructional Resources for General Chemistry,* Washington, D.C., American Chemical Society, 1975.

General References and Popular Accounts; the Chemical Industry

Alexander, G., *Silica and Me — The Career of an Industrial Chemist,* Washington, D. C., American Chemical Society, 1973.

Billmeyer, F. W., Jr., *Entering Industry: A Guide for Young Professionals,* New York, Wiley, 1975.

Cook, G. A., *Survey of Modern Industrial Chemistry,* Ann Arbor, Mich., Ann Arbor Science, 1975.

Encyclopedia of Chemical Technology, New York, Wiley-Interscience, 1978.

Killeffer, D. H., *Chemical Engineering,* Washington, D. C., American Chemical Society, 1975.

McGraw-Hill Encyclopedia of Science and Technology, New York, McGraw-Hill, 1977.

"People in Science," *SciQuest* (a continuing series of articles).

Rochow, E., *Modern Descriptive Chemistry,* Philadelphia, W. B. Saunders, 1977.

Seybold, P. G., *"The Chemical Industry – Employment and Economics,"* Chemistry (October 1978), pp. 14–16.

Vanderbilt, B. M., *Thomas Edison, Chemist,* Washington, D. C., American Chemical Society, 1971.

What's Happening in Chemistry?, Washington, D. C., American Chemical Society (an annual publication).

Witcoff, H., "The Chemical Industry: What Is it?" *Journal of Chemical Education* (April 1979), pp. 253–256.

Woodburn, J. H., *Taking Things Apart and Putting Things Together,* Washington, D. C., American Chemical Society, 1976.

Handbooks and Compilations of Chemical Properties; Laboratory Safety

Green, M. E., *Safety in Working with Chemicals,* New York, Macmillan, 1978.

Handbook of Chemistry and Physics, Cleveland, CRC Press.

Lange's Handbook of Chemistry, New York, McGraw-Hill.

Merck Index, Rahway, N. J., Merck.

Sax, N. I. (ed.), *Dangerous Properties of Industrial Materials,* New York, Van Nostrand Reinhold, 1975.

Steere, N. V. (ed.), *Safety in the Chemical Laboratory,* Easton, Pa., *Journal of Chemical Education,* 1974.

The Toxic Substances List, Washington, D. C., Superintendent of Documents, U.S. Government Printing Office.

Foreign-Language Texts

Consider the translations of the text and lab manual of Masterton and Slowinski:

Chimie: Théorique et Expérimentale, G. S. Gantcheff (translator), Montreal, Les Editions HRW Ltée, 1974.

Experimentelle Einfürhrung in Grundlagen und Methoden der Chemie, D. Krug (translator), Stuttgart, Gustav Fischer Verlag, 1976.

for Sarah and Katharine

Contents

———————— MATTER AND MEASUREMENTS

QUESTIONS TO GUIDE YOUR STUDY

1. What role does chemistry play in today's world? What chemical issues have recently been debated in the news media?

2. What kinds of problems do chemists try to solve? Are there any specially useful or simplifying approaches to their solution?

3. What kinds of materials does the chemist generally work with in the laboratory? Are they the materials of the "real" world, such as steel, plastic, beer, and tobacco smoke?

4. How does the chemist obtain, prepare, and purify these materials?

5. What are some of the instruments and techniques of the chemist?

6. What properties does the chemist measure to describe materials? How are these measurements communicated to other scientists and to the public?

7. What limitations and uncertainties are there in these measurements? How are these communicated?

8. What kind of test would you perform in your kitchen to show that a sample of table sugar is "pure"?

9. What are some of the unsolved problems in chemistry?

10. What questions would you, as a reporter for the campus newspaper, want to ask of a government-employed chemist?

There are surely many other questions you will have as you read the text, listen to lectures, or take part in a discussion of the material of this study. Write them down. Get some answers — from your own work as well as from others involved in the pursuit of chemistry.

YOU WILL NEED TO KNOW

Concepts

Though no previous encounter with chemistry is assumed at this point, at least a general understanding is assumed for some of the basic ideas of chemistry: matter, energy, composition, experiment, measurement, . . .

Math

    1. How to use exponential notation — Appendix 4.*

    2. How to recognize and solve first order (linear) equations, such as y = ax + b. See the Basic Skills section, below, as well as the Selected Readings list at the end of this chapter.

You should also obtain a "scientific calculator," which you will find very helpful in homework, quizzes, and examinations (see Appendix 4).

BASIC SKILLS

In this introductory chapter, you are expected to become familiar with some of the common types of measurements that chemists make and the experiments they carry out to separate and identify substances. You should, for example, be familiar with the measurement of mass and the units in which masses of substances are commonly expressed. Again, you should understand the principles that govern such separation techniques as distillation and fractional crystallization. You should be able to explain how such properties as boiling point and absorption spectrum can be used to identify a substance.

A few specific skills are required in this chapter:

1. Express any number in exponential notation. Use this notation in working calculations.

Throughout your study of chemistry, you will need to use exponential notation. The reason is simple: many of the properties we deal with are either extremely small (for example, masses of individual atoms and molecules) or large (e.g., the numbers of atoms or molecules in any tangible sample of matter). We need a convenient way of expressing such numbers that are much smaller than one or much larger than one.

See Appendix 4 for a discussion of exponential notation. Consider the following example.

- -

    Convert each number to exponential notation. Where a calculation is indicated, work it through and express the answer in this form.

    (a) The mass of a gold atom is about 197/(602 000 000 000 000 000 000 000) g. _____

    197 would be rewritten 1.97×10^2; the second number 6.02×10^{23}.

---

*All chapters and appendices referred to are in *Chemical Principles*. Also see the readings lists for appropriate background and supplementary material.

The quotient becomes: $\dfrac{1.97 \times 10^2}{6.02 \times 10^{23}} = 3.27 \times 10^{-22}$ g.

(b) The average speed of an oxygen molecule in air at 300 K, about room temperature, is about $\left[\dfrac{3 \times 83\ 100\ 000 \times 300}{32.0}\right]^{\frac{1}{2}}$ cm/s. _____

Rewriting: $\left[\dfrac{3 \times 8.31 \times 10^7 \times 3 \times 10^2}{3.20 \times 10}\right]^{\frac{1}{2}} = [\,2.34 \times 10^9\,]^{\frac{1}{2}}$

$$= [(23.4)(10^8)]^{\frac{1}{2}}$$

$$= 4.84 \times 10^4 \text{ cm/s}$$

The last step, showing how to take a square root when the exponent is not an even number, is unnecessary if you use exponential notation on your calculator.

_ _

2. Apply the rules of significant figures to calculations based upon experimental measurements.

These rules are illustrated both in the body of the text and in Examples 1.2 and 1.3. In fact, they are used throughout all calculations in the text and this guide. See Problems 6 and 7 at the end of Chapter 1 in the text.

Significant figures should be used on examinations and in the laboratory too.

3. Use the conversion factor approach to convert measured or calculated quantities from one unit to another.

This is a general method which can be applied to a wide variety of problems. It will be used throughout the text and this guide. If you are not familiar with it, this chapter gives you an opportunity to become skilled in its use.

The principle behind the conversion factor approach is illustrated in Examples 1.4 to 1.6. Its application to conversions involving more than one step is further illustrated here.

_ _

A certain car made in the United States is reported to have a fuel economy of 27.3 mi/gal. Convert this to kilometres per litre.

Two conversions are required: miles to kilometres and gallons to litres. The first conversion factor is found directly in Table 1.1.

(1) 1 mi = 1.609 km

The other factor does not appear directly in the table. However, the conversion may be accomplished by first converting gallons to quarts:

(2)   1 gal = 4 qt

and then to litres:

(3)   1 ℓ = 1.057 qt (U.S.)

With these conversion factors, we can now set up the problem:

$$27.3 \; \frac{mi}{gal} \times \frac{1.609 \; km}{1 \; mi} \times \frac{1 \; gal}{4 \; qt} \times \frac{1.057 \; qt}{1 \ell} = 1.16 \; \frac{km}{\ell}$$

Note that each factor is set up in such a way (1.609 km/1 mi) that the original unit (mi) cancels out, leaving the desired unit (km). That is, initial quantity × conversion factor = desired quantity.

– –

See the listings at the end of the chapter, in the text, of those problems involving this method of converting units. As a general method of solving problems involving units that are unfamiliar to you, consider, for example, Problem 34. Your first step in part (a) might be to convert hides to yards (4 yards = 1 hide). Then go from yards to nookes (4 nookes = 1 yard) and finally to fardells (2 fardells = 1 nooke). The entire conversion could be set up as follows:

$$4.00 \; \text{hides} \times \frac{4 \; \text{yards}}{1 \; \text{hide}} \times \frac{4 \; \text{nookes}}{1 \; \text{yard}} \times \frac{2 \; \text{fardells}}{1 \; \text{nooke}} = 128 \; \text{fardells}$$

4.   **Use an algebraic equation to solve for an unknown quantity, given or having calculated all the other quantities in the equation.**

This is another basic skill which you will use frequently in chemistry. In this chapter, it is illustrated in simple form by

– temperature conversion (°C, °F, K) carried out using Equations 1.2 and 1.3. See Example 1.1 and the exercise directly following it.

– relating density, mass, and volume, using the defining equation

$$\text{density} = \frac{\text{mass}}{\text{volume}} \; ; D = m/V$$

This skill is illustrated by Example 1.3 and the exercise following it. See also the list of problems involving density.

5. **Distinguish between substance and mixture, mixture and solution, and element and compound.**

A descriptive, or nonquantitative, skill such as this is also important for your mastery of the subject. Most or all of the information you need is in the body of the text. The Glossary is also useful in answering questions dealing with the composition of matter.

6. **Describe the separation of mixtures into pure substances and the properties of those substances upon which separation depends.**

Specifically, be able to discuss in a general way the nature of distillation and fractional crystallization. Again, list those properties which would allow you to show that a separation has occurred — properties that identify pure substances.

7. **Use the Periodic Table to locate and distinguish between metals, nonmetals, and metalloids; between main-group and transition elements.** In later chapters, many other uses of the Periodic Table will be described. You are not expected, at this point, to know all of its features, although you may remember many of them from high school.

In the Self-Test on p. 6, as in those for all chapters, there are 20 objective questions (true-false, multiple choice). These are followed by five problems; the last problem is somewhat more difficult than the others. We suggest that you not take this test until you feel that you have a solid grasp of the material in Chapter 1. Then, you should devote one hour (60 min) to taking the test. Pretend it's "for real"; don't refer to your text, notes, etc. When the hour is up, grade yourself as follows:

2 points for each objective question (maximum = 40)

15 points for each of the first four problems (maximum = 60)

10 points for the fifth problem (a bonus if you get it within the hour time limit)

If you follow this system, you should find out how well you understand the material in this chapter. The grade you get should reflect accurately the one you can expect to get on the "real" examination. A high grade (90 or better) suggests that you have a good command of the material in this chapter. If your grade is low, that is a clear signal that you need to spend more time on the material.

SELF-TEST

True or False

1. Changes in physical state, like melting and boiling, tend to   (   )
resolve matter into pure component substances.

2. Physical properties may be used to identify a substance.   (   )

3. In trying to identify a certain liquid compound L, a student   (   )
finds that its density, freezing point, and boiling point are indistin-
guishable from those of a compound Z. The student may safely
assume that L and Z are one and the same.

4. In SI, the units used for pressure and energy are the joule   (   )
and the pascal, respectively.

5. A comfortable room temperature might be about 20°C.   (   )

6. In the Periodic Table, nonmetals are located toward the   (   )
upper right corner.

7. A litre is almost the same as the quart.   (   )

8. In dividng 4.053 by 2.46, your calculator displays the   (   )
answer 1.647 560 976. You should report this number as your
answer.

Multiple Choice

9. The number 0.002 070 should be written in exponential   (   )
notation as
(a) 2.07×10^{-3} (b) 2.070×10^{-3}
(c) 2.070×10^{3} (d) 20.7×10^{-4}

10. How many significant figures are there in the number 6.50   (   )
$\times 10^{3}$?
(a)   two   (b)   three
(c)   four   (d)   six

11. Consider four measurements for the mass of a certain metal   (   )
coin: 3.0 g, 3.006 g, 3.01 g, and 0.003 002 kg. The average mass, in
grams, should be reported as
(a)   3.0045   (b)   3.004
(c)   3.00   (d)   3.0

12. Which one of the following is a transition metal?   (   )
    (a) H   (b) Cl   (c) Cu   (d) Na

13. A procedure appropriate to the separation of the com-   (   )
ponents of gasoline is:
    (a) burning
    (b) fractional crystallization
    (c) fractional distillation
    (d) paper chromatography

14. Which one of the following is a metalloid?   (   )
    (a) Mg   (b) Si   (c) S   (d) Ar

15. A suitable, nondestructive test for determining whether or (   )
not a particular green gem is an emerald would be to:
    (a) determine its melting point
    (b) see if it is scratched by a diamond
    (c) measure its absorption spectrum
    (d) weigh it

16. For a given substance, density ordinarily increases in the (   )
order
    (a) solid, liquid, gas   (b) liquid, solid, gas
    (c) gas, liquid, solid   (d) solid, gas, liquid

17. A steel screw weighs 0.50 oz. Its mass in grams (454 g = 1 lb
= 16 oz) is
    (a) $0.50 \times 454 \times 16$ (b) $0.50 \times 454/16$
    (c) $0.50 \times 16/454$ (d) none of these

18. If the volume and mass measurements on a sample of   (   )
arsenic are 2.10 cm^3 and 12.040 g, the reported value for the density
of arsenic, in g/cm^3 should have how many significant figures?
    (a) two   (b) three
    (c) four   (d) five or more

19. How many times larger is a Celsius degree than a Fahrenheit (   )
degree?
    (a) 1.8   (b) 1/1.8
    (c) 32   (d) they are the same

20. The solid objects A and B are at temperatures of 274 K and (   )
276 K, respectively. When they are placed in contact with each
other,
    (a) heat "flows" from A to B
    (b) heat "flows" from B to A
    (c) the temperature of A drops
    (d) the temperature of B rises

Problems

1.   Suppose that gold is selling for $18.50 per gram. What would be the market value of an ounce of gold at this rate? (Prices in 1979 ended with this value.) 1 g = 0.035 27 oz.

2.   The diameter of a red blood corpuscle is about 8.6×10^3 nm. How many would have to line up side by side to reach across a drop of blood that has a diameter of 0.25 in?   1 in = 2.54 cm; 1 nm = 10^{-7} cm.

3.   The density of mercury is 13.6 g/cm^3 at a certain temperature. What is the volume, in cubic metres, of a sample of mercury weighing 1.00 kg?

4.   Suppose you add 17.9 g of potassium chloride to 100.0 g of water. The resultant solution is found to have a density of 1.104 g/cm^3.
   (a)   What do you calculate for the mass of the solution? (How many significant figures do you get?)
   (b)   What must be the volume of the solution?

5.   Planet X has oceans of liquid ammonia (mp = $-78°$C, bp = $-34°$C). The temperature scale used on Planet X takes the melting point of ammonia to be $0°$X and the boiling point to be $100°$X. That is:

$$
\begin{array}{c c c c}
-34 & - & - & 100 \\
\\
-78 & - & - & 0 \\
\\
°C & & & °X
\end{array}
$$

Express $0°$C in $°$X.

SELF-TEST ANSWERS

1.   **T**   (The basis, for example, of fractional distillation.)
2.   **T**   (These include properties like density, melting point, and solubility.)
3.   **T**   (Though these are only a few clues, identity of properties *must* mean identity of composition.)
4.   **F**   (The order is reversed.)
5.   **T**   (The equivalent Fahrenheit temperature is: 1.8(20) + 32 = 68.)
6.   **T**
7.   **T**   (You will find it helpful to *know* at least one conversion factor relating English and metric units of distance, volume, and mass.)

8. **F** (Round off to the correct number of significant figures! Report 1.65 instead.)

9. **b** (As written, the number implies an uncertainty in the last place: 0.002 070 ± 0.000 001. The zero at the right should be retained.)

10. **b** (The exponential notation shows that the 6.50 contains these digits.)

11. **d** (The least accurate result is known only to within 0.1 g, so a calculated average of 3.0045 would be rounded to the nearest 0.1.)

12. **c** (You need to use a Periodic Table to be sure.)

13. **c** (The bulk of the mixture consists of liquid substances.)

14. **b**

15. **c**

16. **c** (Water is exceptional. See Chapter 11.)

17. **b** (Check for unit cancellation.)

18. **b**

19. **a** (See Figure 1.4.)

20. **b** (The temperature of A rises, that of B drops.)

Solutions to Problems

1. Setting this up so as to convert from the given unit (oz) to the desired unit, its equivalent in dollars ($):

$$1 \text{ oz} \times \frac{1 \text{ g}}{0.035\ 27 \text{ oz}} \times \frac{\$18.50}{1 \text{ g}} = \$524.5$$

2. $$0.25 \text{ in} \times \frac{2.54 \text{ cm}}{1 \text{ in}} \times \frac{1 \text{ nm}}{10^{-7} \text{ cm}} \times \frac{1 \text{ corpuscle}}{8.6 \times 10^3 \text{ nm}}$$
$$= 7.4 \times 10^2 \text{ corpuscles}$$

3. $$1.00 \text{ kg} \times \frac{10^3 \text{ g}}{1 \text{ kg}} \times \frac{1 \text{ cm}^3}{13.6 \text{ g}} \times \frac{1 \text{ m}^3}{10^6 \text{ cm}^3} = 7.35 \times 10^{-5} \text{ m}^3$$

4. (a) 17.9 g + 100.0 g = 117.9 g (Four; both masses are known to ± 0.1 g)

   (b) Rearranging the equation, density = mass/volume:

   $$\text{volume} = \frac{\text{mass}}{\text{density}} = \frac{117.9 \text{ g}}{1.104 \text{ g/cm}^3} = 106.8 \text{ cm}^3$$

5. $°X = a(°C) + b$ 　　　　　$100 = -34a + b$
   　　　　　　　　　　　　　$0 = -78a + b$
   $a = 100/44 = 2.27$ 　　　　　$b = 177$
   $°X = 2.27(°C) + 177$ 　　　so: $0°C = 177°X$

SELECTED READINGS

Analytical tools of the chemist are considered in:

Alexander, G., *Chromatography: An Adventure in Graduate School,* Washington, D.C., American Chemical Society, 1977.
Davis, J.C., Introduction to Spectroscopy, *Chemistry* (October 1974), pp. 6–10.
Keller, R.A., Gas Chromatography, *Scientific American* (October 1961), pp. 58–67.
Storms, H.A., Probing Concentration Zero, *Chemistry* (March 1973), pp. 6–10.
Walker, J., A Homemade Spectrophotometer Scans the Spectrum in a Thirtieth of a Second, *Scientific American* (January 1980), pp. 150–160.

Chemistry past, present, and future; careers in chemistry:

Adcock, L.H., Chemistry 200 Years Ago, *Chemistry* (September 1975), pp. 14–15.
Chemistry in the 1980's, *Chemical and Engineering News* (November 26, 1979), pp. 29–59.
Facts and Figures for the Chemical Industry, *Chemical and Engineering News* (an annual feature).
Hill, B.W., Careers in the Chemical Industry, *Chemistry* (November 1975), pp. 6–9.
What's Happening in Chemistry? Washington, D.C., American Chemical Society (an annual publication).
Witcoff, H., The Chemical Industry: What Is It? *Journal of Chemical Education* (April 1979), pp. 253–256.

Chemistry safe and unsafe:

Brown, M.H., *Laying Waste: The Poisoning of America by Toxic Chemicals,* New York, Pantheon, 1980.
Green, M.E., *Safety in Working with Chemicals,* New York, Macmillan, 1978.
Lehmann, P.E., The High Price of Art, *SciQuest* (September 1979), pp. 13–16.
Meselson, M., Chemical Warfare and Chemical Disarmament, *Scientific American* (April 1980), pp. 38–47.
Wheeler, G., Chemical Hazards in the Arts, *Journal of Chemical Education* (April 1980), pp. 281–282.
Zentner, R.D., Hazards in the Chemical Industry, *Chemical and Engineering News* (November 5, 1979), pp. 25–34.

For more on the Periodic Table, see Chapter 2 and its readings list.

For more on problem solving, see the readings listed in the Preface.

_____ ATOMS, MOLECULES, AND IONS

QUESTIONS TO GUIDE YOUR STUDY

1. Can you think of common, everyday observations which suggest that matter is made of atoms and molecules? Or that it isn't?

2. Why are some of these building blocks of matter neutral, while others are charged?

3. How would you experimentally show that sulfur and oxygen combine in a one-to-one mass ratio to form the gaseous compound sulfur dioxide? How would you demonstrate the conservation of mass?

4. Just how small are atoms and molecules? Is there a convenient way of counting them and of expressing their numbers? Of weighing them?

5. If more than one kind of atom exists for a particular element, then how are we to interpret the atomic mass of the element?

6. What holds atoms together in a molecule?

7. What are the supporting arguments, based on observation, for the atomic theory?

8. How do you distinguish, in terms of atomic theory, between an element and a compound? Or, between two different compounds of carbon and oxygen?

9. If the idea of atoms is at least as old as ancient Greece, then why did it take so long for the atomic theorists to get anywhere?

10. If not an atom, then what is the fundamental, ultimate building-block?

YOU WILL NEED TO KNOW

Concepts

1. That a chemical symbol may represent one or more atoms of an element; a molecular formula, one or more molecules of a substance — Chapter 1 (Section 1.3). See this and the next chapters also.

2. The distinction between element and compound — Chapter 1 (Section 1.3). In fact, you should recognize several elements and compounds by name and by formula. This will come with repeated use.

Math

No other math skill is required for this chapter than that which you have already used for Chapter 1.

BASIC SKILLS

1. **Use mass composition data for two or more compounds of two elements to illustrate the Law of Multiple Proportions.** More generally, explain the laws of chemistry in terms of atomic theory.

Example 2.1 and the exercise following it illustrate the Law of Multiple Proportions. To illustrate the Law of Constant Composition, consider the following example.

When sulfur dioxide is formed from the elements, sulfur and oxygen combine in a one-to-one mass ratio. If 2.25 g of sulfur are combined with oxygen to form the compound, then what are the masses of oxygen? _____ of sulfur dioxide? _____

The masses of sulfur and oxygen must compare 1:1. So, 2.25 g S combine with 2.25 g O. The Law of Conservation of Mass is illustrated by the fact that there must be 2.25 + 2.25 = 4.50 g sulfur dioxide formed.

The importance of these laws and the calculations based on them cannot be overemphasized. See the catalog of problems at the end of the chapter for further illustration. Chapters 3 and 4 provide many more examples of calculations based primarily on the laws of constant composition and conservation of mass.

2. **Relate the numbers of protons and neutrons in an atom to its nuclear symbol.**

The nuclear symbol gives the atomic number at the lower left of the symbol and the mass number at the upper left.

atomic number = number of protons

mass number = number of protons + number of neutrons

This skill is shown in Example 2.2.

Note that the atomic number locates the element in the Periodic Table. This number also gives the number of electrons in a neutral atom of the element.

3. Relate the numbers of electrons and protons to the charge on a monatomic ion.

See Example 2.3 and the catalog of problems at the end of the chapter. Note that in a positive ion (cation), one or more electrons have been removed to give an excess of protons over electrons and so a net positive charge. A negative ion (anion) is formed when one or more electrons are added to a neutral atom.

$$\text{charge} = (\text{no. of protons}) (+1) + (\text{no. of electrons}) (-1)$$

$$= \text{no. of protons} - \text{no. of electrons}$$

4. Relate the charges on anion and cation to the formula for an ionic compound.

This skill is illustrated in the body of the text and is given more attention in later chapters. Note that the sum of the charges on all ions in the formula for the compound must be zero.

— —

The ionic compound magnesium chloride has the formula $MgCl_2$. If the anion is Cl^-, what must be the charge on the cation? _____ What formula would be written for magnesium oxide, where the oxide ion is O^{2-}? _____
Electrical neutrality requires that

$$\text{charge on 1 Mg} + \text{charge on 2 Cl} = 0$$

$$\text{charge on Mg} = 0 - 2(-1) = +2$$

The formula for the oxide, MgO, is the result of the "cancellation" of one +2 charge by one -2 charge.

— —

5. Relate the atomic mass of an element to the abundances and masses of its isotopes.

Example 2.4 and the exercise following it require the calculation of atomic mass from the isotopic abundances and masses. The reverse calculation is illustrated in the following example.

- -

Thallium (atomic mass = 204.37) consists of two isotopes, ^{203}Tl (mass = 203.05) and ^{205}Tl (mass = 205.05). What is the abundance (%) of the heavier isotope? _____

Let us represent the abundance of the heavy isotope, ^{205}Tl, by x. Since the abundances of the two isotopes must add up to 100%, the abundance of the light isotope must be 100 – x. Equation 2.3 becomes:

$$\text{atomic mass Tl} = 204.37 = \frac{x}{100}(205.05) + \frac{(100-x)}{100}(203.05)$$

Multiplying both sides of the equation by 100:

$$20\ 437 = 205.05x + 20\ 305 - 203.05x$$

$$x = 132/2.00 = 66.0\%$$

- -

6. Given the formula of a compound, determine its formula mass.

This skill is illustrated for molecular substances by Example 2.5 and the exercise following it. Note that when it is known that a substance is molecular, this mass is generally called the molecular mass. Consider now an ionic substance.

- -

Determine the formula mass of common salt, NaCl. _____ of calcium carbonate, $CaCO_3$. _____

Analogous to the calculation of a molecular mass:

formula mass = sum of atomic masses of all atoms in formula

So, for NaCl:

$$\text{formula mass} = \text{atomic mass Na} + \text{atomic mass Cl}$$

$$= 22.99 + 35.45 = 58.44$$

For $CaCO_3$:

$$\text{formula mass} = 40.08 + 12.01 + 3(16.00) = 100.09$$

- -

Note that a subscript in a formula applies to the symbol written directly before it. In $CaCO_3$, there are three O atoms, one C, and one Ca.

   7. **Relate the numbers of atoms (or molecules) and the mass in grams of a sample of matter.**

This skill is based on an understanding of Avogadro's number and is illustrated by Example 2.6 and the exercise following it. See the catalog of problems for additional examples. Note that in using this skill, you must be familiar with exponential notation. The numbers you calculate are often either very large or very small.

   8. **Given the formula of a substance, relate the number of moles and the mass in grams of a sample.**

The conversion between moles and grams is a very important one. It is illustrated in Example 2.7 and below, in combination with Skill 7.

- -

   Given 12.0 g H_2O (molecular mass = 18.0), calculate the number of moles. _____ and the number of molecules. _____
   The necessary "conversion factors" are:

$$1 \text{ mol } H_2O = 6.022 \times 10^{23} \text{ molecules } H_2O$$

$$= 18.0 \text{ g } H_2O$$

$$\text{no. moles} = 12.0 \text{ g} \times \frac{1 \text{ mol}}{18.0 \text{ g}} = 0.667 \text{ mol}$$

$$\text{no. molecules} = 12.0 \text{ g} \times \frac{6.022 \times 10^{23} \text{ molecules}}{18.0 \text{ g}} = 4.01 \times 10^{23} \text{ molecules}$$

- -

You will have ample opportunity to apply this last skill — throughout the remainder of the course! In particular, you will very often be required to convert from moles to grams or vice versa. Remember that, although a mole always contains a definite number of items (6.022×10^{23}), its mass in grams varies from one species to another. Thus a mole of H_2O weighs 18.0 g, a mole of NaCl weighs 58.44 g, and so on.

In addition to these quantitative skills, there are a few descriptive skills you should become familiar with.

— Describe the components of an atom with respect to charge and mass.

— Distinguish between structural and molecular formulas. (For example, compare H—Cl and HCl.) More discussion of what formulas represent comes in the next and later chapters.

— Describe experimental methods of determining atomic masses.

SELF-TEST

True or False

1. The chemical properties of an atom are determined by its nuclear charge. ( )

2. An isotope is one of two or more atomic species having the same atomic number but different numbers of electrons. ( )

3. The molecular mass of a substance is simply a number which tells how heavy a molecule is when compared to a chosen reference. ( )

4. All the metals have large mass numbers. ( )

5. All neutral atoms of a given element have the same number of electrons. ( )

6. Stable, bulk samples of matter carry little or no electric charge, that is, they are electrically neutral. ( )

7. The atomic mass of chlorine is about 35.5. Chlorine consists of two isotopes of mass numbers 35 and 37. It follows that the lighter isotope must be the more abundant. ( )

Multiple Choice

8. For the ion, $^{40}_{20}Ca^{2+}$, how many electrons surround the nucleus? ( )
    (a) 18   (b) 20
    (c) 22   (d) some other number

9. Which of the following has the smallest mass? ( )
    (a) an atom of C   (b) a molecule of CO_2
    (c) one gram of C   (d) one mole of C

10. A certain isotope X has an atomic number of 7 and a mass   (   )
number of 15. Hence,
    (a) X is an isotope of nitrogen
    (b) X has eight neutrons per atom
    (c) an atom of X has seven electrons
    (d) all of the above

11. To illustrate the Law of Multiple Proportions, we could use   (   )
data giving the percentages by mass of
    (a) Mg in MgO and in $MgCl_2$
    (b) S in Na_2SO_3 and in Na_2SO_4
    (c) glucose in two different mixtures with sucrose
    (d) $^{63}_{29}Cu$ and $^{65}_{29}Cu$ isotopes in copper metal

12. Suppose the atomic mass scale had been set up with calcium   (   )
chosen for a mass of exactly 10 units, rather than about 40 on the
present scale. On such a scale, the atomic mass of oxygen would be
about
    (a) 64   (b) 32
    (c) 16   (d) 4

13. A sample of an element is   (   )
    (a) a collection of atoms with identical number of neutrons
    (b) a collection of atoms with identical nuclear masses
    (c) a collection of atoms with identical nuclear charges
    (d) one of the following: air, earth, fire, water

14. Experimental support for the existence of atoms and   (   )
subatomic particles includes all of the following except
    (a) electrical currents through gases
    (b) metal foil scattering of alpha particles
    (c) radioactive decay
    (d) continuous mechanical subdivision of a single crystal

15. There is always a ratio of small whole numbers between   (   )
    (a) the atomic and molecular masses of a diatomic element
    (b) the mass percentages of Cu in any two of its compounds
    (c) the masses of Cu combined with 1 g of A in CuA and
       1 g of B in CuB
    (d) the atomic masses of any two elements

16. The mass of an individual atom is of the order of   (   )
    (a) 10^{-22} kg   (b) 1/2000 g
    (c) 10^{-22} g   (d) 6×10^{23} g

17. One atom of He (atomic mass = 4.00) weighs $6.63 \times$ (   )
10^{-24} g. One atom of oxygen (atomic mass = 16.0) must weigh
    (a) 16.0 g   (b) 6.02×10^{-23} g
    (c) 2.66×10^{-23} g   (d) 1.66×10^{-24} g

18. The idea that most of the mass of an atom is concentrated   (   )
in a very small core, the nucleus, is a result of the experiments of
    (a) Dalton   (b) Dulong and Petit
    (c) Rutherford   (d) Thomson

19. In one mole of hydrogen, H_2 there are Avogadro's number   (   )
of
    (a) H atoms   (b) electrons
    (c) H_2 molecules   (d) neutrons

20. A mole of acetic acid, $C_2H_4O_2$ (AM C = 12, H = 1, O = 16,   (   )
    (a) contains N molecules, where N is Avogadro's number
    (b) contains 8 N atoms
    (c) weighs 60 g
    (d) all of the above

Problems

1. The atomic mass of gallium is 69.72.
    (a) How many times heavier is an atom of Ga than an atom of
        $^{12}_{6}C$?
    (b) If there are two isotopes, with masses 68.96 and 70.96, what
        are their abundances (%)?

2. A molecule of octane contains 8 carbon and 18 hydrogen atoms.
    (a) What is the molecular mass of octane? (AM C = 12.0, H = 1.0)
    (b) How many molecules are there in a sample of octane weighing
        26.0 g?
    (c) How many moles are there in such a sample?

3. The density of iron (AM = 55.8) is 7.86 g/cm^3. Calculate the
volume, in cubic centimetres, occupied by one
    (a) mole of iron
    (b) atom of iron

4. Phosphoric acid has the molecular formula H_3PO_4 (AM H = 1.0, P
= 31.0, O = 16.0).
    (a) What is the mass in grams of 1.43 mol of H_3PO_4?
    (b) How many moles of H_2O are required to have the same mass
        as 1.43 mol of H_3PO_4?

5. The radius of the He nucleus is about 1×10^{-14} m. Estimate the
density of the helium nucleus in grams per cubic centimeter (AM He = 4.0).
The volume of a sphere is $4\pi r^3/3$.

SELF-TEST ANSWERS

1. **T** (This is the number of protons and also the atomic number.)
2. **F** (Different numbers of neutrons.)
3. **T** (That reference is carbon-12.)
4. **F** (Lithium, for example, has mass number 7 in its most abundant isotope. Metals lie to the left of the zigzag line — Chapter 1.)
5. **T**
6. **T** (This principle is applied when writing the formula of an ionic compound; see Basic Skill 4.)
7. **T** (Can you explain this conclusion?)
8. **a** (Two electrons have been removed, leaving two units more of positive charge than of negative charge in the atom; 20 protons and 18 electrons.)
9. **a** (In order of increasing mass: $a < b < c < d$.)
10. **d** (c applies to a neutral atom.)
11. **b** (The same elements, two, or in this case three, must be present in the compounds compared.)
12. **d** (The masses of the individual atoms are still in the same ratio: 4/10 is the same as 16/40.)
13. **c** (Compare to Question 1.)
14. **d** (Division could not continue indefinitely. You would ultimately come up against identical units of structure: atoms, molecules, or ions.)
15. **a** (Molecules contain whole numbers of atoms. Consider H and H_2.)
16. **c** (Divide any molar mass of an element by Avogadro's number, 6.022×10^{23}.)
17. **c** (An atom of oxygen is 16.0/4.00, or four times heavier.)
18. **c**
19. **c** (One mole of items, here H_2 molecules, is the same as N items.)
20. **d** (Notice that each molecule contains 8 atoms.)

Solutions to Problems

1. (a) $\dfrac{\text{mass of Ga atom}}{\text{mass of }^{12}\text{C atom}} = \dfrac{69.72}{12} = 5.810$

(b) Letting x be the abundance of the lighter isotope, substitution into Equation 2.3 gives:

$$69.72 = \frac{x}{100}(68.96) + \frac{(100 - x)}{100}(70.96)$$

$$x = 62.0$$

$$100 - x = 38.0$$

2. (a) $8(12.0) + 18(1.0) = 114$

(b) $26.0 \text{ g} \times \dfrac{6.022 \times 10^{23} \text{ molecules}}{114 \text{ g}} = 1.37 \times 10^{23}$ molecules

(c) $26.0 \text{ g} \times \dfrac{1 \text{ mol}}{114 \text{ g}} = 0.228$ mol

3. (a) $1.00 \text{ mol} \times \dfrac{55.8 \text{ g}}{\text{mol}} \times \dfrac{1 \text{ cm}^3}{7.86 \text{ g}} = 7.10 \text{ cm}^3$

(b) $1 \text{ atom} \times \dfrac{7.10 \text{ cm}^3}{6.022 \times 10^{23} \text{ atoms}} = 1.18 \times 10^{-23} \text{ cm}^3$

4. (a) $1.43 \text{ mol} \times \dfrac{98.0 \text{ g}}{1 \text{ mol}} = 140 \text{ g}$

(b) $140 \text{ g H}_2\text{O} \times \dfrac{1 \text{ mol}}{18.0 \text{ g H}_2\text{O}} = 7.78 \text{ mol}$

5. $\text{mass} = \dfrac{4.0}{6.0 \times 10^{23}} \text{ g} = 7 \times 10^{-24} \text{ g}$

$\text{volume} = \dfrac{4\pi}{3}(1 \times 10^{-12} \text{ cm})^3 = 4 \times 10^{-36} \text{ cm}^3$

$\text{density} = \dfrac{\text{mass}}{\text{volume}} = \dfrac{7 \times 10^{-24} \text{ g}}{4 \times 10^{-36} \text{ cm}^3} = 2 \times 10^{12} \text{ g/cm}^3$

SELECTED READINGS

Atomic theory and its history are considered in:

Feinberg, G., Ordinary Matter, *Scientific American* (May 1967), pp. 126–134.
Lagowski, J.J., *The Structure of Atoms,* Boston, Houghton Mifflin, 1964.
Lucretius, *On the Nature of the Universe,* Baltimore, Penguin, 1951.
Nash, L.K., *Stoichiometry,* Reading, Mass., Addison-Wesley, 1966.
Patterson, E.C., *John Dalton and the Atomic Theory,* Garden City, N.Y., Doubleday, 1970.

Avogadro's number and its determination, and the mole:

Hildebrand, J.H., *Principles of Chemistry,* New York, Macmillan, 1964.
Kieffer, W.F., *The Mole Concept in Chemistry,* New York, Van Nostrand, 1973.

Esoteric researches of high- and low-energy physics are the subject matter of many recent issues of Scientific American, Chapter 24, and:

Golub, R., Ultracold Neutrons, *Scientific American* (June 1979), pp. 134–154.
Thomson, G.P., *J.J. Thomson: Discoverer of the Electron,* Garden City, N.Y., Doubleday, 1966.

The Periodic Table is considered also in the readings of Chapters 5, 9 and 12:

Jaffe, B., *Moseley and the Numbering of the Elements,* Garden City, N.Y., Doubleday, 1971.
Kragh, H., On the Discovery of Element 72, *Journal of Chemical Education* (July 1979), pp. 456–459.
Puddephatt, R.J., *The Periodic Table of the Elements,* New York, Oxford, 1972.
Rochow, E.G., *Modern Descriptive Chemistry,* Philadelphia, W.B. Saunders, 1977.
Sanderson, R.T., *Chemical Periodicity,* New York, Reinhold, 1960.
Seaborg, G.T., Prospects for Further Considerable Extension of the Periodic Table, *Journal of Chemical Education* (October 1969), pp. 626–634.
Weeks, M.E., *Discovery of the Elements,* Easton, Pa., Journal of Chemical Education, 1968.

CHEMICAL FORMULAS

QUESTIONS TO GUIDE YOUR STUDY

1. What information is conveyed by a chemical formula? What may a formula represent, in terms of the number of particles, moles, or mass?

2. You have seen formulas written for atoms, molecules, and ions. How are they different from one another?

3. What are some of the experimental procedures used to determine chemical formulas?

4. How would you experimentally establish the percent composition by mass for a given compound? How would you relate this to the relative numbers of atoms that combine?

5. How close to ratios of small whole numbers of atoms do you need to come before you round off to such whole numbers?

6. What minimum information is required to write a formula? What assumptions, if any, need to be made?

7. What kind of experiment would you do to show that the formula of water is H_2O and not HO as Dalton believed?

8. What regularities exist in the formulas for the elements and their positions in the Periodic Table?

9. How are substances named? What is the relationship between the formula for a compound and its name?

10. Can you predict the formula for the combination of two given elements, perhaps based on their positions in the Periodic Table?

YOU WILL NEED TO KNOW

Concepts

1. How to interpret formulas — in particular, the meaning and use of subscripts. Additional work is done on this topic in the current chapter. Symbols for the elements were first used in Section 1.3.

Math

1. How to work problems using the conversion factor method. This approach began with Chapter 1 and continues throughout. We refer to it here because it is very useful beginning with the chemical calculations of this chapter and the next.

2. How to relate a chemical formula, formula mass, number of moles, and the mass of a sample of any given substance − Chapter 2. Be able to calculate one from another. See Skills 7 and 8, Chapter 2 of this guide.

3. How to write the formula for a simple ionic compound given the formulas for cation and anion − Chapter 2. This is an application of the principle of electrical neutrality. See Basic Skill 4 in Chapter 2 of this guide.

BASIC SKILLS

Stoichiometry is an awesome label usually attached to the arithmetic of chemical reactions. It is the quantitative application of chemical formulas and equations, the shorthand notation that represents the conservation of mass and constant composition. The practical importance of stoichiometry is easily demonstrated − the principles established in this and the next chapter will be used throughout the text as well as in any laboratory work you do.

This chemical arthmetic is not particularly difficult or complicated. Its mastery may, however, require considerable practice in the following skills. Perhaps even more important is the fundamental Skill 8 of Chapter 2.

1. Write and interpret the formulas for some common substances.

This skill is illustrated by several substances discussed in the first section of the chapter. These substances include elements (see Table 3.1), molecular compounds, ionic compounds, and hydrates. Additional examples are seen throughout the chapter. Interpretation of a chemical formula includes the following ideas.

- -

What is meant by each of the following formulas?

(a) H_2 _____ (b) K_2CO_3 _____ (c) $CaCl_2 \cdot 2H_2O$ _____

The fact that the element hydrogen has the formula H_2 indicates that it consists of diatomic molecules rather than isolated H atoms. When the formula H_2 appears in a chemical equation, we ordinarily take it to represent one mole of the element.

The formula $K_2 CO_3$ represents the ionic compound potassium carbonate. (Naming the compound is considered in a later section of the chapter and in a skill below.) If the formula is taken to represent one mole of substance, then it also represents 2 mol K^+ ions and 1 mol CO_3^{2-} ions. Again, it gives the simplest ratio of whole numbers of atoms found combined in the compound (2 K: 1 C: 3 O).

The formula $CaCl_2 \cdot 2 H_2O$ indicates the composition of a hydrate of calcium chloride, $CaCl_2$. The notation "$\cdot 2 H_2 O$" tells us that there are two moles of water per mole of $CaCl_2$ in the hydrate. In turn, the ionic compound calcium chloride contains 1 mol of Ca^{2+} ions and 2 mol of Cl^- ions per mole of $CaCl_2$.

_ _

2. Given the formula of a compound, calculate the percentages by mass of the elements.

The skill is illustrated by Examples 3.1 and 3.2 and the exercises that follow them. Also see the catalog of problems at the end of the chapter. A slightly different view of this skill is seen in the following problem, and is not unlike the second part of Example 3.2.

_ _

The important iron ore magnetite has the formula Fe_3O_4. How much of the metal could one hope to extract from a hundred grams of the ore?_____
The mass of iron in a 100 g sample has the same numerical value as the percentage by mass. The percent of Fe is the same regardless of the sample size. So, for example, let us choose a 1 mol sample:

$$\text{mass of 1 mol } Fe_3O_4 = \text{mass of 3 mol Fe} + \text{mass of 4 mol O}$$

$$= 3(55.8) + 4(16.0), \text{ in units of grams}$$

where 55.8 and 16.0 are the atomic masses of Fe and O.

$$\% \text{ Fe} = \frac{\text{mass of 3 mol Fe}}{\text{mass of 1 mol } Fe_3O_4} \times 100$$

$$= \frac{3(55.8)}{3(55.8) + 4(16.0)} \times 100 = 72.3$$

So, in 100 g sample:

$$\text{g Fe} = 100 \text{ g } Fe_3O_4 \times \frac{72.3 \text{ g Fe}}{100 \text{ g } Fe_3O_4} = 72.3$$

_ _

Note that if we were asked to determine the amount of Fe in x grams of ore, the final expression would look like this: g Fe = $(x)\left(\dfrac{72.3 \text{ g Fe}}{100 \text{ g Fe}_3\text{O}_4}\right)$.

3. Determine the simplest formula of a compound, given the mass percentages of the elements or analytical data from which these can be calculated.

This skill is used in Examples 3.3 and 3.4 and their accompanying exercises, where the mass percentages are directly given. The procedure followed can be outlined stepwise as follows.

– Calculate the numbers of moles of the elements present in 100 g of the compound. (Note that the number of grams equals the percent of each element in such a sample.)

– Determine the mole ratio. This is numerically equal to the atom ratio.

– Determine the atom ratio (small whole numbers).

Analytical data is used to calculate the percentage composition in Example 3.5 and the exercise that follows it. Note that the percent composition and the ratio of masses that combine do not depend on the units of mass one employs. Do not be tricked by the mg unit in the example.

4. Determine the molecular formula of a compound, given the simplest formula and at least an approximate molecular mass.

Example 3.6 shows how this calculation is made. You compare the molecular and simplest formula masses in order to find an integer, a small whole number by which to multiply all the subscripts in the simplest formula. See the catalog of problems for additional examples.

5. Use the Periodic Table to obtain the charges of ions formed by the main-group elements.

See the discussion at the beginning of Section 3.5. You will be expected to know that elements in Groups 1, 2, 6, and 7 form ions with charges of +1, +2, -2, and -1 in that order.

6. Write the formula for an ionic compound given either the formulas of the ions or the name of the compound.

The first skill is illustrated by Example 3.7. You should recognize this as a straightforward application of the principle of electrical neutrality discussed in Chapter 2. Also see Skill 4 in Chapter 2 of this guide. Example 3.8 extends this skill by using the data summarized in Figure 3.3 and Table 3.2.

Writing the formula to fit the name of an ionic compound is illustrated in the exercises following these examples. More discussion of names follows in Section 3.6 of the text. Review the rules for naming simple ionic compounds.

7. Given the formula for a compound, give its name.

This is done for simple ionic compounds, binary compounds of nonmetallic elements, and for several common (nonsystematically named) compounds. Examples using this skill include 3.9 and 3.10.

8. Describe some experimental methods of determining percent composition.

Like the first skill listed in this chapter, this is basically a nonquantitative, descriptive skill. It is important to recognize some of the experimental background for the science. See Section 3.3 in the text.

SELF-TEST

True or False

1. Since three out of every four atoms in a sample of $ScCl_3$ are ( )
chlorine atoms, 75% of the mass of the sample is due to chlorine.

2. The % by mass of iron is greater in $FeCl_2$ than in $FeCl_3$.   ( )

3. A molecular formula may be written only for a compound. ( )

4. The molecular formula of hydrogen peroxide is H_2O_2. This ( )
is also the simplest formula of hydrogen peroxide.

5. In each mole of $(NH_4)_2CO_3$, there are three moles of ions. ( )

6. From chemical analysis one finds that 1.801 mol of Cl ( )
combine with 0.9017 mol of Ca. With the atom ratio of Cl to Ca
being $1.801/0.9017 = 1.997$, one should give the formula of calcium
chloride as $CaCl_{1.997}$ and not $CaCl_2$.

7. A certain compound containing only silver, nitrogen, and ( )
oxygen is known to be 63.5% Ag and 8.2% N by mass. Unless an
experiment is done to determine the % oxygen, the simplest formula
cannot be obtained.

Multiple Choice

8. The simplest formula of a substance always shows ( )
   - (a) the elements present and the simplest ratio of whole numbers of atoms
   - (b) the actual numbers of atoms combined in a molecule of the substance
   - (c) the number of molecules in a sample of the substance
   - (d) the molecular mass of the substance

9. The formula for calcium carbonate, $CaCO_3$ (formula mass = ( )
100), generally represents all of the following except
   - (a) one formula unit of calcium carbonate
   - (b) N formula units of calcium carbonate, where N is Avogadro's number
   - (c) one gram of calcium carbonate
   - (d) one hundred grams of calcium carbonate

10. A compound with the simplest formula C_2H_5O has a ( )
molecular mass of 90 (AM C = 12, H = 1, O = 16). The molecular
formula for the compound is:
   - (a) $C_3H_6O_3$
   - (b) $C_4H_{26}O$
   - (c) $C_4H_{10}O_2$
   - (d) $C_5H_{14}O$

11. In analyzing a compound for carbon, the compound is ( )
burned in air and the masses of products determined. What
assumption do we make?
   - (a) all the oxygen in the air is converted to water, H_2O
   - (b) all the carbon is converted to carbon dioxide, CO_2
   - (c) equal numbers of moles of CO_2 and H_2O are produced
   - (d) all the oxygen in the H_2O produced comes from the compound

12. From analytical data, you calculate a mole ratio of 0.40 P : ( )
1.00 O. How would you write the simplest formula for this
compound of phosphorus and oxygen?
   - (a) $P_{0.40}O$
   - (b) PO_2
   - (c) P_2O_5
   - (d) P_5O_2

13. Potassium chromate, K_2CrO_4, contains 40.3% K, 26.8 % Cr ( )
by mass. The mass percent of oxygen must be:
   - (a) 40.3 + 26.8 = 67.1
   - (b) 100.0 – 67.1 = 32.9
   - (c) 4(16.0) = 64.0
   - (d) determined by experiment

14. Which contains the largest number of molecules? (Molecular ( )
masses are shown in parentheses.)
     (a) 1.0 g CH_4 (16)   (b) 1.0 g H_2O (18)
     (c) 1.0 g HNO_3 (63)   (d) 1.0 g N_2O_4 (92)

15. About how much oxygen is there in exactly one mole of ( )
baking soda, $NaHCO_3$?
     (a) 16 g   (b) 24 g
     (c) 48 g   (d) 96 g

16. The maximum amount of silver one might hope to recover ( )
from a kilogram of waste silver chloride that is only 85% AgCl would
be, in kilograms (formula masses: Ag = 108, AgCl = 143),
     (a) 108/143   (b) 0.85 X 108/143
     (c) 0.85 X 143/108   (d) 0.85 X 108/143 X 1000

17. Which of the following formulas represents copper sulfate ( )
pentahydrate?
     (a) CuS   (b) $CuSO_4$
     (c) $CuSO_4 + 5\ H_2O$ (d) $CuSO_4 \cdot 5\ H_2O$

18. The ionic compound containing Fe^{3+} and $SO_4{}^{2-}$ should ( )
have the formula
     (a) $FeSO_4$ (b) Fe_2SO_4
     (c) $Fe_2(SO_4)_3$ (d) $Fe_3(SO_4)_2$

19. The compound of the preceding question would be called ( )
     (a) ferrous sulfate   (b) iron(II) sulfate
     (c) iron(III) sulfate   (d) iron(III) sulfide

20. How would the molecular compound N_2O_3 be named? ( )
     (a) nitrogen oxide   (b) nitrogen(III) oxide
     (c) dinitrogen trioxide   (d) ammonia

Problems

1. Complete the following table:

        Formula   Name

     (a) NH_3 _____
     (b) Na_2CO_3 _____
     (c) $FeSO_4$ _____
     (d) _____   lithium oxide
     (e) _____   ammonium nitrate
     (f) _____   iron(III) hydroxide

2. Calculate the mass percent composition of potassium dichromate, $K_2Cr_2O_7$. (atomic masses: K = 39.1, Cr = 52.0, O = 16.0)

3. A 0.600 g sample of pure tin (atomic mass Sn = 119) is reacted completely with gaseous fluorine (atomic mass F = 19.0) to form 0.984 g of solid tin fluoride. What is the simplest formula of the tin fluoride?

4. When 3.42 g of a compound containing only the elements K, Cl, and O is heated, it decomposes to give 2.08 g KCl and 1.34 g O_2. (atomic masses: K = 39.1, Cl = 35.5, O = 16.0)

    (a) Calculate the masses of K and Cl present in the original compound.

    (b) Calculate the number of moles of each element in the original compound.

    (c) Determine the simplest formula of the compound.

5. Analysis of a gaseous mixture of NO and N_2O shows it to contain 52% by mass nitrogen. What is the % by mass of NO in the mixture? (atomic masses: N = 14, O = 16)

SELF-TEST ANSWERS

1. **F** (Atom for atom, Sc is heavier than Cl; actual % Cl = 70.)
2. **T** (Try to give a simple explanation.)
3. **F** (Examples of elements include O_2 and P_4. Ionic compounds would not have molecular formulas.)
4. **F** (The simplest formula is HO.)
5. **T** (The structural units you should recognize as NH_4^+ and CO_3^{2-}. See Table 3.2 in the text.)
6. **F** (Besides the possibility that the data may not be even this good in an actual analysis, what reasons can you give for rounding off?)
7. **F** (You would calculate % O by difference. The three percents must add up to 100.)
8. **a** (The information in b and d would be obtained from the molecular formula.)
9. **c**
10. **c** (This is the only formula with the correct atom ratio.)
11. **b** (See the discussion preceding Example 3.5 in the text.)
12. **c** (This is the same as 1:2.5 or, finally, 2:5.)
13. **b** (See Question 7 above.)
14. **a** (The largest fraction of a mole, 1.0/16. This is a Chapter 2 review question.)

15. c   (Three moles of the element, O.)
16. b
17. d
18. c   (Total charge must be zero.)
19. c   (The numeral refers to the charge on the cation. $SO_4{}^{2-}$ is called sulfate.)
20. c   (See Section 3.6 in the text; nitrogen and oxygen are both nonmetals.)

Solutions to Problems

1. (a) ammonia   (b) sodium carbonate   (c) iron(II) sulfate
   (d) Li_2O (e) NH_4NO_3 (f) $Fe(OH)_3$
2. Formula mass $K_2Cr_2O_7 = 2(39.1) + 2(52.0) + 7(16.0) = 294$

$$\% \ K = 100\% \times \frac{2 \times \text{atomic mass K}}{\text{formula mass } K_2Cr_2O_7} = \frac{100 \times 2(39.1)}{294} = 26.6$$

    $\%Cr = 100 \times 2(52.0)/294 = 35.4$
    $\%O = 100 \times 7(16.0)/294 = 38.1$ (or, by difference, $100 - 26.6 - 35.4$)
3. moles Sn = 0.600 g $\times$ 1 mol/119 g = 0.005 04
   moles F = (0.984 – 0.600) g $\times$ 1 mol/19.0 g = 0.0202
   atom ratio = mole ratio = 0.0202 F/0.005 04 Sn = 4 F/1 Sn
   simplest formula: SnF_4 (Write the symbol of the metal first, the nonmetal last.)
4. (a) g K in original compound = g K in 2.08 g KCl (formula mass = 74.6)
                              = 2.08 g KCl $\times$ 39.1 g K/74.6 g KCl
                              = 1.09
    g Cl = 2.08 – 1.09 = 0.99
  (b) moles O = 1.34 g $\times$ 1 mol/16.0 g = 0.0838
     moles K = 1.09 g $\times$ 1 mol/39.1 g = 0.0279
     moles Cl = 0.99 g $\times$ 1 mol/35.5 g = 0.0279
  (c) Dividing each number of moles by 0.0279 gives:
     moles O:moles K:moles Cl = 3:1:1
     formula: $KClO_3$
5. In 100 g mixture, there are X g NO:

$$52 \ g \ N = \frac{14 \ g \ N}{30 \ g \ NO} \ (X \ g \ NO) + \frac{28 \ g \ N}{44 \ g \ N_2O} \ (100 - X)g \ N_2O$$

    X = 71 g; 71% NO

SELECTED READINGS

Chemical formulas, among other topics, are considered in:

Copley, G.N., Linear Algebra of Chemical Formulas and Equations, *Chemistry* (October 1968), pp. 22–27.
Kieffer, W.F., *The Mole Concept in Chemistry*, New York, D. Van Nostrand, 1973.
Soltzberg, L., *BASIC and Chemistry*, Boston, Houghton Mifflin, 1975.

The readings of Chapter 4 also deal with formulas and extend the discussion to equations. Also see the problem-solving texts listed in the Preface.

4

CHEMICAL REACTIONS AND EQUATIONS

QUESTIONS TO GUIDE YOUR STUDY

1. How do you distinguish between *reaction* and *equation?*
2. What information is conveyed by a chemical equation?
3. What does it mean to "balance" an equation?
4. What may a chemical equation represent in terms of the number of particles, moles, or mass of reactants and products? Just how do you quantitatively relate amounts of substances reacting and forming?
5. How would you experimentally establish the equation for a reaction?
6. How do you represent the physical state (solid, liquid, solution . . .) of the materials taking part in a reaction?
7. Are the conditions, such as temperature and pressure, under which a reaction occurs represented by the equation for the reaction?
8. How do you write equations for precipitation, acid-base, and oxidation-reduction reactions?
9. What does a chemical equation tell you about what you would see as a reaction occurs?
10. What does an equation tell you about what the molecules are doing as a reaction proceeds?

YOU WILL NEED TO KNOW

Concepts

1. Formulas for some common substances — see Table 3.1 (molecules formed by the elements) and Table 3.2 and Figure 3.3 (common ions) in the text. The more familiar you are with these formulas and names, the easier will be the task of writing equations.

Math

1. How to solve an algebraic equation for an unknown quantity, given all other quantities in the equation — review Skill 4 in Chapter 1 of this guide.

2. How to calculate one of the quantities — number of moles, grams, or particles — from another, given the formula for a substance — see Skills 7 and 8 of Chapter 2 of this guide.

BASIC SKILLS

1. Write and balance simple chemical equations.

This skill is described in Section 4.1 of the text where the balanced equation is derived for the reaction of SnO_2 with H_2 to give tin and water vapor. Example 4.1 and the exercise that follows it further illustrate this skill. In each case, the procedure consists of:

— writing an unbalanced equation. This skeletal equation shows the formulas of all reactants and products. You may have to look up some of these at first (see Tables 3.1 and 3.2 and Figure 3.3 in the text). Eventually, you will be expected to know the formulas and usual physical states of such common substances as carbon dioxide, oxygen, and ammonia.

— balancing the equation. Whenever possible, begin with an element which appears in only one reactant and one product. Adjust only coefficients, never subscripts.

— labeling the physical states of all reactants and products.

Additional examples include the equations of Section 4.5, representing reactions of ions in water solution. Note that here one new factor is involved in balancing equations. The total charge must be the same on both sides. For example, the equation

$$Zn(s) + Ag^+(aq) \rightarrow Zn^{2+}(aq) + Ag(s)$$

is *not* balanced even though there are the same number of atoms of each element on both sides. In the balanced equation there is a charge of +2 on both sides:

$$Zn(s) + 2\ Ag^+(aq) \rightarrow Zn^{2+}(aq) + 2\ Ag(s)$$

More practice in balancing equations of this type in a systematic way comes in later chapters.

2. Given a balanced equation, relate the numbers of moles of any two substances taking part in the reaction. Compare the numbers of individual particles involved in the reaction.

The underlying principle here is: the coefficients of a balanced equation can represent *either* the relative numbers of moles, *or* the number of particles, of reactants and products. These coefficients can be used to derive conversion factors which show the chemical equivalence of any two substances or species involved in the reaction. This application of the conversion factor method of working problems is shown in Example 4.2 and the exercise following it.

This mole/mole relation is fundamental to all other calculations involving chemical equations. *It must be mastered here and now.* See the catalog of problems at the end of the chapter for practice in this skill.

3. Given the balanced equation, relate the masses of any two substances taking part in a reaction. Relate the number of moles of one substance to the number of grams of another.

Review Skill 8 of Chapter 2 in this guide. You need to be able to calculate the mass corresponding to a number of moles of any substance, given the formula of the substance. Examples 4.3 and 4.4 illustrate this skill in detail. Notice again the consistent use of conversion factors here, as in previous chapters and in the skill above. Experience convinces us that this approach is the most general and reliable way of analyzing a wide variety of problems in chemistry. If you are still using the "ratio and proportion" method, stop. Most of the problems that you will see in homework, on exams, or in the text simply cannot be worked by this rote approach. Besides, it will be helpful if we are using the same language.

See the catalog of problems at the end of the chapter for additional practice in this skill.

4. Given or having calculated two of the three quantities – concentration (M), number of moles of solute, volume of solution – calculate the other quantity.

The application of this skill involves two others you have seen: Skill 4 of Chapter 1 and Skill 8 of Chapter 2. The defining relation is Equation 4.3: M = moles solute/litres (or dm^3) solution. Examples 4.5 and 4.6 illustrate the skill in detail.

An extension of the skill is seen in Example 4.7. Given the concentration of reagent used to prepare a solution, you should be able to calculate the concentration of each solute species. A solution of an ionic compound is assumed to involve complete dissociation of the compound to its component cation and anion. So again, another skill is included: you need to recognize common ions. (Example: the ions expected to be released in a solution of $Al_2(SO_4)_3$ are Al^{3+} and SO_4^{2-}.) See Table 3.2 and Figure 3.3.

   5. **Given the balanced equation for a reaction involving species in solution, relate the volumes or concentrations of two reactant species.**

This skill is illustrated in Examples 4.8 to 4.10; see the catalog of problems for additional practice. Two simple principles are involved:
   (a) The coefficients of the equation, as always, relate the numbers of moles of species taking part in the reaction (Skill 2).
   (b) The number of moles of a species in solution is given by the relation:

$$\text{no. of moles} = M \times \text{volume (in litres or } dm^3)$$

Thus, for a reaction: $A(aq) + 2\ B(aq) \rightarrow$ products

we could say:

$$\text{no. moles B} = 2 \times (\text{no. moles A})$$
$$(M\ B) \times (\text{volume B}) = 2 \times (M\ A) \times (\text{volume A})$$

   6. **Describe reactions in water solution as involving precipitation, acid-base, and oxidation-reduction.**

More extensive discussion of these classes of reactions comes in later chapters. For now, at least be able to correctly label such reactions and give definitions for such terms as acid, precipitate, and oxidation.

SELF-TEST

True or False

   1. As used in chemistry, *equation* and *reaction* are interchange- (   )
able.

   2. An equation is balanced when the total number of atoms of (   )
each element is the same on either side of the equation and the charges add up to being the same on either side.

3. In any chemical reaction, the number of moles of reactants ( )
equals the number of moles of products.

4. That species which loses one or more electrons in a reaction ( )
is said to be reduced.

5. Precipitation refers to the formation of a liquid product.   ( )

6. The volume of a solution in litres (or dm³) can be found by ( )
multiplying the number of moles by the concentration in moles per
litre (or mol/dm³).

7. The concentration of a solution of an ionic compound is ( )
equal to the sum of the concentrations of cation and anion.

Multiple Choice

8. To write a chemical equation for a given reaction, you need ( )
to know at least the
    (a)   masses of all reactants and products
    (b)   mole ratios of all reactants and products
    (c)   formulas of all reactants and products
    (d)   all the above

9. Ammonia and oxygen can be made to react to form only ( )
nitrogen and liquid water. The reactants are:
    (a) NH_3 and O   (b) NH_3 and O_2
    (c) NH_3 and O_3 (d) N_2 and H_2O

10. For the reaction just considered, the balanced equation is: ( )
    (a) $2\ NH_3(g) + 3\ O(g) \rightarrow N_2(g) + 3\ H_2O(1)$
    (b) $4\ NH_3(g) + 3\ O_2(g) \rightarrow 2\ N_2(g) + 6\ H_2O(1)$
    (c) $2\ NH_3(g) + O_3(g) \rightarrow N_2(g) + 3\ H_2O(1)$
    (d) $2\ N_2(g) + 6\ H_2O(1) \rightarrow 4\ NH_3(g) + 3\ O_2(g)$

11. Again consider the reaction of ammonia and oxygen. The ( )
number of moles of oxygen consumed for each mole of nitrogen
formed is:
    (a) 0.67   (b) 1.5
    (c) 2.0   (d) 3.0

12. Which set of numbers are the coefficients for the balanced ( )
equation for the combustion of octane?
$$\_\_\_C_8H_{18}(1) + \_\_\_O_2(g) \rightarrow \_\_\_CO_2(g) + \_\_\_H_2O(1)$$

    (a) 1, 25, 8, 18   (b) 1, 25/2, 16, 18
    (c) 1, 25, 8, 9   (d) 2, 25, 16, 18

13. If 4.5 mol of NO_2 are allowed to react with excess H_2O ( )
according to the equation

$$3 NO_2(g) + H_2O(1) \rightarrow 2 HNO_3(1) + NO(g)$$

the yield of HNO_3, in moles, would be

(a) 3/2 X 4.5   (b) 4.5
(c) 2/3 X 4.5   (d) 2.0

14. Baking soda ($NaHCO_3$, formula mass = 84.0) and hydro- ( )
chloric acid (HCl, formula mass = 36.5) react according to the
equation

$$HCO_3^-(aq) + H^+(aq) \rightarrow H_2O(1) + CO_2(g)$$

How many grams of baking soda would be consumed in forming
10.0 g of CO_2 (formula mass = 44.0)?

(a) $\dfrac{10.0}{44.0}$ (b) $\dfrac{10.0 \times 36.5}{44.0}$

(c) $\dfrac{10.0 \times 84.0}{44.0}$ (d) $\dfrac{44.0 \times 84.0}{10.0}$

15. The number of moles of solute in 500 cm^3 of a 1.6 M ( )
solution is:

(a) 500 X 1.6   (b) $\dfrac{500}{1000}$ X 1.6

(c) $\dfrac{500}{1000 \times 1.6}$ (d) $\dfrac{1.6}{500}$

16. To determine the concentration of $H^+(aq)$ in a water ( )
solution of HBr, one would probably carry out

(a) a precipitation   (b) an oxidation-reduction
(c) a distillation   (d) an acid-base titration

17. What is the concentration of anion in 0.10 M $Al(NO_3)_3$? ( )
(Assume complete dissociation.)

(a) 0.10 M   (b) 0.30 M
(c) 0.40 M   (d) it depends on the
                          sample volume

18. To obtain 0.015 mol of Al^{3+} from 0.10 M $Al(NO_3)_3$, you ( )
would measure out _____ cm^3 of solution.

(a) 0.0015   (b) 0.015
(c) 0.15   (d) 150

19. Listed below are the labels on four bottles of 1 M water ( )
solutions. In which case would the total concentration of ions be the
greatest?

(a) NaCl   (b) $FeSO_4$
(b) Na_2SO_4 (d) $FeCl_3$

20. The balanced equation for the reaction of phosphine with   (   )
air is: $4\ PH_3\ (g) + 8\ O_2\ (g) \rightarrow 6\ H_2O(1) + P_4O_{10}(s)$.

    34   32   18   284

Molecular masses have been written below the formulas. Which of
the following statements apply to this reaction?
- (a) For each molecule of P_4O_{10} formed, four molecules of PH_3 disappear.
- (b) For each mole of P_4O_{10} formed, twelve moles of reactants are consumed.
- (c) For each mole of P_4O_{10} formed, $6(18)$ g H_2O also form.
- (d) All of the above.

Problems

1. Write a balanced chemical equation for each of the following reactions.
- (a) At 300 K and atmospheric pressure, water is decomposed to the elements.
- (b) A precipitate forms on mixing solutions containing $Pb^{2+}(aq)$ and $Cl^-(aq)$.
- (c) Acetylene, $C_2H_2\ (g)$, is burned in air to give carbon dioxide and liquid water.

2. Consider the reaction of *one* mole of H_2S according to the equation $2\ H_2S(g) + 3\ O_2\ (g) \rightarrow 2\ SO_2\ (g) + 2\ H_2O(1)$. Atomic masses are: $H = 1.0, O = 16.0, S = 32.1$.
- (a) How many moles of water are formed?
- (b) How many grams of sulfur dioxide are formed?

3. A solution is prepared by dissolving 12.0 g of K_2CrO_4 in enough water to make 2.00×10^2 cm^3 of solution (AM $K = 39.1$, Cr $= 52.0$, O $= 16.0$). Calculate the concentration of:
- (a) K_2CrO_4 (b) K^+ (c) $CrO_4{}^{2-}$

4. What volume (cm^3) of 0.100 M HCl would be required to react with 0.250 g of zinc (atomic mass $= 65.38$) if the equation for the reaction is $Zn(s) + 2\ H^+(aq) \rightarrow Zn^{2+}(aq) + H_2\ (g)$?

5. A sample of pure iron(III) oxide weighing 0.120 g is heated. The only products are oxygen and another oxide of iron, having a mass of 0.116 g. (AM Fe $= 55.8$, O $= 16.0$)
- (a) How many moles of iron are present in the first oxide? the second oxide?

    (b) What is the simplest formula of the second oxide?

    (c) Write the balanced equation for the reaction.

SELF-TEST ANSWERS

1. **F** (An equation represents a reaction.)
2. **T**
3. **F** (There is a conservation of mass, not moles.)
4. **F** (Oxidized.)
5. **F** (Solid product, coming out of solution.)
6. **F** (Volume = no. of moles/M.)
7. **F** (The concentration is given in terms of the reagent used to prepare the solution and not the species that may form as a result of dissociation.)
8. **c** (Note that the information in a and b can be obtained from this choice once the equation is balanced.)
9. **b** (You should be familiar with formulas of several common substances. See Skill 6 of Chapter 3 of this guide and Tables 3.1 and 3.2 and Figure 3.3 of Chapter 3 in the text.)
10. **b** (Reactants are written to the left of the arrow.)
11. **b** (Using the coefficients of the equation to relate the numbers of moles: $3 \text{ mol } O_2/2 \text{ mol } N_2$.)
12. **d**
13. **c** (Using the coefficients as a conversion factor: $4.5 \text{ mol } NO_2 \times 2 \text{ mol } HNO_3/3 \text{ mol } NO_2$.)
14. **c** (To check this answer, attach appropriate units.)
15. **b**
16. **d**
17. **b** (The solute species are Al^{3+} and NO_3^-; see Table 3.2 and Figure 3.3 of the text.)
18. **d** (From the definition of M: litres (dm^3) of solution = moles solute/M = 0.015/0.10 = 0.15; 150 cm^3.)
19. **d** (Consider complete dissociation: $FeCl_3(s) \rightarrow Fe^{3+}(aq) + 3 Cl^-(aq)$.)
20. **d** (You should be able to relate numbers of particles, moles, and masses for any balanced equation.)

Solutions to Problems

1. (a) $2 H_2O(l) \rightarrow 2 H_2(g) + O_2(g)$
    (b) $Pb^{2+}(aq) + 2 Cl^-(aq) \rightarrow PbCl_2(s)$ (Note charge balance.)
    (c) $2 C_2H_2(g) + 5 O_2(g) \rightarrow 4 CO_2(g) + 2 H_2O(l)$

2. (a) moles H_2O = 1 mol $H_2S \times$ 2 mol H_2O/2 mol H_2S = 1 mol
   (b) grams SO_2 = 1 mol $H_2S \times$ 2 mol SO_2/2 mol $H_2S \times$ 64.1 g/mol
        SO_2 = 64.1 g
3. (a) FM K_2CrO_4 = 2(39.1) + 52.0 +4(16.0) = 194.2

   $$M\ K_2CrO_4 = \frac{12.0/194.2\ \text{mol}}{0.200} \frac{}{\ell} = 0.309\ M$$

   (b) M K^+ = 2 $\times$ M K_2CrO_4 = 0.618
   (c) M $CrO_4{}^{2-}$ = M K_2CrO_4 = 0.309

4. The volume we are interested in must supply the number of moles of solute H^+(aq) that are equivalent to the number of moles of Zn given. So first convert grams of Zn to moles; then relate moles of Zn to moles of H^+. From there we can calculate volume of solution using volume solution (in litres or dm^3) = moles solute/M. Phew! The conversion factors are easier to follow.

$$\text{litres solution} = 0.250\ \text{g Zn} \times \frac{1\ \text{mol Zn}}{65.38\ \text{g}} \times \frac{2\ \text{mol H}^+}{\text{mol Zn}} \times$$

$$\frac{1\ \ell\ \text{solution}}{0.100\ \text{mol H}^+}$$

$$= 0.0765\ \ell = 0.0765\ \text{dm}^3$$

volume solution = 76.5 cm^3

5. (a) Moles Fe in second oxide = moles Fe in first oxide (Fe_2O_3)

   $$\text{moles Fe} = 0.120\ \text{g Fe}_2O_3 \times \frac{1\ \text{mol Fe}_2O_3}{159.6\ \text{g Fe}_2O_3} \times \frac{2\ \text{mol Fe}}{\text{mol Fe}_2O_3}$$

   $$= 1.50 \times 10^{-3}$$

   (b) Note that since we have calculated the amount of Fe in the second oxide, we can now calculate the amount of O, by difference.
   grams Fe = 1.50×10^{-3} mol $\times$ 55.8 g/mol = 0.084 g
   grams O = 0.116 – 0.084 = 0.032 g
   moles O = 0.032 g $\times$ 1 mol/16.0 g = 2.0×10^{-3}
   So, the mole ratio is 2 O:1.5 Fe, or 4 O:3 Fe; the formula, Fe_3O_4.
   (c) 6 Fe_2O_3(s) → 4 Fe_3O_4(s) + O_2(g)

SELECTED READINGS

Problem solving requires practice. For a selection of exercises, work as many of the problems in the text as you can. Also, consider one or more of the problem manuals listed in the Preface to this guide. And, particularly for balancing equations, consider:

Copley, G.N., Linear Algebra of Chemical Formulas and Equations, *Chemistry* (October 1968), pp. 22–27.
Greene, D.G.S., An Algebraic Method for Balancing Chemical Equations, *Chemistry* (March 1975), pp. 19–21.

Kieffer, W.F., *The Mole Concept in Chemistry,* New York, D. Van Nostrand, 1973.
Strong, L.E., Balancing Chemical Equations, *Chemistry* (January 1974), pp. 13–15.

The use of computers in chemistry is discussed in:

Soltzberg, L., *BASIC and Chemistry,* Boston, Houghton Mifflin, 1975.

Additional readings on reactions in water solution can be found listed for Chapters 17, 18, and 21.

5

───────── SOURCES OF THE ELEMENTS

QUESTIONS TO GUIDE YOUR STUDY

1. What are the relative abundances of the elements at the earth's surface? Which elements are most abundant, which the least? What geological processes lead to the observed distribution of elements?

2. Which elements occur uncombined? Which are too reactive to be found uncombined? Is there a correlation with position in the Periodic Table?

3. Of those elements occurring in compounds, which usually occur as simple cations or anions?

4. Which metals occur as oxides, which as sulfides? How do the metalloids occur in nature?

5. Given a particular element of commercial value, say a metal, can you describe its principal source, the method of extraction, and it uses?

6. What reactions (e.g., acid-base, oxidation-reduction, precipitation) are employed in metallurgy? What equations would you write?

7. How much energy is involved in the overall extraction process? What are the costs, directly, in money; what are they in long-term environmental and health effects? How many people are employed in such industrial chemistry?

8. How is a nonmetal such as hydrogen prepared commercially? How would you prepare a sample of H_2 in the laboratory?

9. What would happen if you used nonstoichiometric (non-equivalent) amounts of reactants in carrying out any of these reactions?

10. What are the limits to the natural reserves of elements? How long will they last at the present rate of consumption? What are the alternatives to their exhaustion?

YOU WILL NEED TO KNOW

This chapter applies the principles of the preceding chapters, with special emphasis on the concepts and calculations introduced in Chapters 3 and 4. Review this section of each of the preceding chapters in this guide to see what background is assumed.

BASIC SKILLS

This chapter completes a unit of study by reviewing and applying all the major ideas and skills of Chapters 1 to 4. Except for the first two skills listed below, there are no new kinds of calculations called for in the examples and problems of this chapter. Perhaps your best preparation for this chapter is to read over the skills stated in Chapters 3 and 4 of this guide. Where you are unsure of yourself, work some more problems involving that skill.

Specifically, the skills reviewed include:

— Given the formula of a compound, calculate the percentage by mass of an element in the compound (Example 5.1)

— Given a balanced equation, relate the numbers of moles of two substances taking part in the reaction (Example 5.2)

— Relate moles solute, volume of solution, and concentration (Examples 5.4, 5.6)

— Determine the simplest formula from mass percent composition (Example 5.5)

— Write and balance a chemical equation (Example 5.7)

— Relate the masses of two substances taking part in a reaction (Example 5.7).

New skills introduced in this chapter include the following.

1. Given the number of moles or masses of all reactants, determine which is the limiting reactant and calculate the theoretical yield of product.

This skill is illustrated by Example 5.2 and the exercise following it. Again, it is applied in working Examples 5.4(b) and 5.8. Perhaps a nonchemical example will further illustrate the reasoning involved.

— —

A bartender has available a litre bottle of gin and another litre bottle of vermouth. How many litres of martinis (4 parts gin to 1 part vermouth) can he prepare? _____

Clearly, the limiting reactant is the gin; to use all the vermouth would require that four litres of gin be available rather than one. If the bartender uses all the gin, he will need only ¼ ℓ of vermouth and will get ⁵/₄ ℓ of martinis. The "equation" for the process is:

$$1 \text{ ℓ gin} + ¼ \text{ ℓ vermouth} \rightarrow ⁵/₄ \text{ ℓ martinis}$$

— —

For more sober applications of this skill, see the catalog of problems at the end of the text chapter.

2. **Calculate the percent yield, given the actual and theoretical yields.**

These three quantities are related by the defining equation (5.2)

$$\% \text{ yield} = \frac{\text{actual yield}}{\text{theoretical yield}} \times 100\%$$

If any two of these three quantities are known, the third can be calculated. See the discussion following Example 5.2. The skill is shown in Example 5.3 and below.

— —

A student carrying out a reaction gets 12.0 g of product. This represents 52% of the yield he should have obtained if the limiting reactant were completely consumed and the product were completely recovered. What must be the theoretical yield? _____

Solving the above equation for theoretical yield:

$$\text{theoretical yield} = \text{actual yield} \times \frac{100\%}{\% \text{ yield}}$$

$$= 12.0 \text{ g} \times \frac{100}{52} = 23 \text{ g}$$

— —

3. **Describe the sources, reactions involved in extraction, and the uses for each of several elements.**

There is a considerable amount of descriptive chemistry with which you should become familiar. Specifically, this skill is illustrated in the text by the discussion of these elements:

N_2, O_2, Ar, H_2, S, the halogens, Na, Mg, Al, Cu, Fe

You should know where these elements are found in nature: the atmosphere, seawater, or minerals on or beneath the surface of the earth. For elements obtained from mineral deposits, you should be familiar with the type of ore involved (free element, oxide, sulfide, and so on). Moreover, you should know what reactions are involved in extracting the elements from their ores.

Besides describing such reactions as are used, you should be able to write balanced equations for them and state some of the reaction conditions. Note whether a given reaction is acid-base or precipitation. Describe the electrolytic preparations and the other oxidation-reduction reactions as well.

Like it or not, to become proficient in this area may require some memorization of factual material. Organize some of this material by using the Periodic Table (e.g., to relate the principal ores to the position of the element in the table).

SELF-TEST

True or False

1. The halogens are all members of the same period in the ( )
Periodic Table.

2. The calculation of theoretical yield is based on the ( )
assumption that all of the limiting reactant is consumed according to the equation for the reaction.

3. One kilogram of the light metal Mg weighs less than 1 kg of ( )
the heavy metal Pb.

4. In an electrolytic cell, reduction occurs at the anode.   ( )

5. The principal source of magnesium is seawater.   ( )

6. At the current rate of growth of consumption, it seems ( )
likely that in less than a century we will exhaust most commercial metals.

Multiple Choice

7. At the earth's surface, the two most abundant substances in ( )
the atmosphere are
    (a) CO_2, H_2O (b) N_2, O_2
    (c) N_2, CO_2 (d) $SiO_2, NaCl$

8. The separation of nitrogen from its principal source involves ( )
    (a) filtration   (b) chromatography
    (c) fractional crystallization   (d) fractional distillation

9. Commercially, hydrogen is usually obtained from   ( )
    (a) the atmosphere   (b) water
    (c) natural gas   (d) the hydrogen bomb

10. Which of the elements listed below commonly occurs as -1 ( )
anions?

(a) H   (b) Na
(c) S   (d) Cl

11. Metals toward the left of a transition series such as Cr and ( )
Mn are likely to be found in nature as

(a) the free metal   (b) carbonates
(c) oxides   (d) sulfides

12. $CaCO_3$ is used in the preparation of iron in the blast furnace ( )
so as to

(a) remove oxides of Si, Mn, and P
(b) remove carbon as CO_2
(c) increase the calcium content
(d) form a protective glassy surface for the pig iron

13. Choose an important natural source of the element copper: ( )

(a) Cu   (b) $CuSO_4 \cdot 5\ H_2O$
(c) $CuCl_2$ (d) Cu_2S

14. What is the oxidation (anode) reaction in the electrolytic ( )
preparation of Mg?

(a) $Mg \rightarrow Mg^{2+} + 2\ e^-$
(b) $Mg^{2+} + 2\ e^- \rightarrow Mg$
(c) $2\ Cl^- \rightarrow Cl_2 + 2\ e^-$
(d) $MgCl_2\ (1) \rightarrow Mg(s) + Cl_2\ (g)$

15. Per mole of compound (formula masses in parentheses), the ( )
largest mass of Mg might be recovered from

(a) $MgCl_2$ (95)   (b) $MgCO_3 \cdot CaCO_3$ (184)
(c) $Mg(OH)_2$ (58)   (d) all contain the same
                                 amount

16. Of the halogens, the one most easily reduced is ( )

(a) F_2 (b) Cl_2
(c) Br_2 (d) I_2

17. A metal sulfide is converted to the metal or an oxide by the ( )
process of

(a) electrolysis   (b) flotation
(c) metallurgy   (d) heating in air

18. In carrying out the reaction at elevated temperatures ( )

$$2\ NaHCO_3\ (s) \rightarrow Na_2CO_3\ (s) + H_2O(g) + CO_2\ (g)$$

a student obtains an 80% yield. How many moles of $NaHCO_3$ must
she have started with if her yield was 1.6 mol Na_2CO_3?

(a) 4.0   (b) 3.2
(c) 2.6   (d) 2.0

19. The actual yield is usually less than 100% because   (   )
    (a) separation and purification result in losses
    (b) competing reactions form other products
    (c) not all the limiting reactant is consumed even when reaction has ceased
    (d) all the above

20. When an excess of one reactant is used   (   )
    (a) more product may form than with no excess
    (b) reaction may proceed at a faster rate
    (c) less limiting reactant may remain unconsumed
    (d) all the above

Problems

1. Write balanced equations for the reactions:
    (a) coke reacts with carbon dioxide to form carbon monoxide
    (b) iron(III) oxide is reduced to the metal by reaction with CO
    (c) aluminum is prepared by electrolysis of the metal oxide (show the electrode reactions and the overall reaction)
    (d) magnesium hydroxide is brought into solution by reaction with hydrochloric acid

2. Chloride is present in seawater at 0.58 M. How many grams of NaCl would you dissolve in water to give 100 cm^3 of solution that had the same concentration of Cl$^-$? (Formula mass NaCl = 58.4)

3. High-grade iron ore may contain as much as 70% by mass of Fe_2O_3. How many grams of iron can be obtained from one kilogram of this ore? (atomic masses: Fe = 55.8, O = 16.0)

4. Hydrogen is readily prepared in the laboratory by the reaction of zinc with excess acid. The metal is oxidized to Zn^{2+}(aq).
    (a) Write the balanced equation for the reaction
    (b) If hydrogen is to be prepared from 0.25 g Zn, what volume of 0.10 M HCl must be used if a 10% excess of acid is called for? (atomic masses: H = 1.0, Cl = 35.5, Zn = 65.4)
    (c) If 3.3 × 10^{-3} mol H_2 is actually recovered, what is the % yield?

5. A certain sample of pig iron contains 2.26% C and 0.24% P. In making steel, O_2 is added to lower the percentage of C to 1.00 by oxidation to CO_2 and remove all the P by oxidation to P_4O_{10}. How many moles of O_2 should be used per kilogram of pig iron? (atomic masses: C = 12.0, P = 31.0, O = 16.0)

SELF-TEST ANSWERS

1. **F** (They are members of the same vertical column, the same *group* or family.)
2. **T**
3. **F** (The masses are the same; the densities are different.)
4. **F** (Oxidation occurs at the anode, reduction at the cathode.)
5. **T**
6. **T** (See Table 5.4.)
7. **b** (See Table 5.1 in the text.)
8. **d** (Air is liquefied, then distilled. See Section 5.1 discussion.)
9. **c** (Can you write the equation (5.1 in the text) for the reaction?)
10. **d** (The same is true for other halogens.)
11. **c** (See the discussion at the beginning of Section 5.4).
12. **a**
13. **d** (Some of the metal also occurs uncombined. The sulfide is expected for a metal at the right of a transition series.)
14. **c** (When molten $MgCl_2$ is electrolyzed, Mg^{2+} is reduced while Cl^- is oxidized. The equation in (d) represents the overall reaction.)
15. **d** (There is 1 mol Mg in 1 mol of each compound. On a per gram basis, what would your answer be?)
16. **a** (The most reactive.)
17. **d** (A process called roasting.)
18. **a** (For 80% yield, every 2.0 mol $NaHCO_3$ gives 0.8 mol Na_2CO_3.)
19. **d** (For explanation of c, see Chapter 13.)
20. **d** (The chapters on reaction rate and chemical equilibrium take this up in detail.)

Solutions to Problems

1. (a) $C(s) + CO_2(g) \rightarrow 2\ CO(g)$
   (b) $Fe_2O_3(s) + 3\ CO(g) \rightarrow 2\ Fe(s) + 3\ CO_2(g)$ (Reaction 5.19)
   (c) $4\ Al^{3+} + 12\ e^- \rightarrow 4\ Al$ (cathode)

   $\underline{6\ O^{2-} \rightarrow 3\ O_2 + 12\ e^-}$ (anode)

   $2\ Al_2O_3(1) \rightarrow 4\ Al(1) + 3\ O_2(g)$ (Reaction 5.14)
   (d) $Mg(OH)_2(s) + 2\ H^+(aq) \rightarrow Mg^{2+}(aq) + 2\ H_2O$ (Reaction 5.11)
2. moles solute $= 0.58 \times 0.100 = 0.058$
   grams solute = moles $\times$ g/mol $= 0.058$ mol $\times$ 58.4 g/mol
   $= 3.4$ g
3. 700 g $Fe_2O_3 \times \dfrac{111.6 \text{ g Fe}}{159.6 \text{ g } Fe_2O_3} = 4.9 \times 10^2$ g Fe

4. (a) $Zn(s) + 2 H^+(aq) \rightarrow Zn^{2+}(aq) + H_2(g)$

   (b) Zn is the limiting reactant, so we calculate

   moles Zn = 0.25 g × 1 mol/65.4 g = 3.82×10^{-3} mol Zn

   moles H^+ required (theoretically) = 3.82×10^{-3} mol Zn ×
   $$\frac{2 \text{ mol } H^+}{\text{mol Zn}}$$
   $$= 7.64 \times 10^{-3} \text{ mol } H^+$$

   for 10% excess, moles H^+ = 110% × 7.64×10^{-3} mol
   $$= 8.40 \times 10^{-3}$$

   volume solution = moles solute/M = $8.40 \times 10^{-3}/0.10$
   $$= 0.084 \ell = 0.084 \text{ dm}^3$$

   (c) theoretical yield = 3.82×10^{-3} mol Zn × 1 mol H_2/mol Zn
   $$= 3.82 \times 10^{-3} \text{ mol } H_2$$
   $$\% \text{ yield} = \frac{3.3 \times 10^{-3}}{3.8 \times 10^{-3}} \times 100 = 87$$

5. oxidation of C: $12.6 \text{ g C} \times \dfrac{1 \text{ mol } O_2}{12.0 \text{ g C}} = 1.05 \text{ mol } O_2$

   oxidation of P: $2.4 \text{ g P} \times \dfrac{5 \text{ mol } O_2}{124 \text{ g P}} = 0.10 \text{ mol } O_2$

   total moles O_2 = 1.05 + 0.10 = 1.15

SELECTED READINGS

For additional practice and reading on problem solving, see the manuals listed in the Preface as well as the book by Kieffer listed in Chapter 3.

On the limits to elemental availability and to the growth of science and technology:

Lapp, R.E., *The Logarithmic Century,* Englewood Cliffs, N.J., Prentice-Hall, 1973.
Meadows, D.H., et al., *The Limits to Growth,* New York, Universe, 1974.
Price, D.J. de S., *Little Science, Big Science,* New York, Columbia, 1963.
Skinner, B.J., A Second Iron Age Ahead?, *American Scientist* May–June 1976), pp. 258–269.

Sources and reactions of the elements, including industrial processes:

Bachmann, H.G., The Origin of Ores, *Scientific American* (June 1960), pp. 146–156.
Cook, G.A., *Survey of Modern Industrial Chemistry,* Ann Arbor, Mich., Ann Arbor Science, 1975.
Reilly, J.J., Hydrogen Storage in Metal Hydrides, *Scientific American* (February 1980), pp. 118–129.
Rochow, E.G., *Modern Descriptive Chemistry,* Philadelphia, W.B. Saunders, 1977.
Sellers, N., Chemistry of Steel Making. *Journal of Chemical Education* (February 1980), pp. 139–142.

6

THERMOCHEMISTRY

QUESTIONS TO GUIDE YOUR STUDY

1. How many chemical reactions going on in the world around us can you think of in which the energy change plays an important role? (What "drives" an automobile engine? A mountain climber?)

2. What kinds of energy may be transferred during chemical reactions?

3. What is the source of the energy involved in a reaction? What happens to it? Can this energy be understood in terms of what happens to the atoms?

4. Is the energy change of a reaction quantitatively related to the reacting masses? (Can you cite a *simple* example?)

5. How is information concerning energy change expressed within the chemical equation? Does your interpretation of the energy change depend on how you write the equation for a reaction?

6. Does the energy change depend on the conditions under which a reaction is carried out? If so, how?

7. How do you measure the energy change for any given reaction? Can it be calculated for a reaction without ever actually carrying out that particular reaction?

8. Are there regularities observed in the energy changes for given kinds of reactions? For example, is the heat flow the same for all acid-base reactions? Can correlations be made with the Periodic Table?

9. What do thermochemical principles allow you to say about practical problems, such as the relative merits of two different fuels?

10. What are some of the fundamental problems in supplying the energy needed by modern society today and tomorrow? Are there known limitations or perhaps untapped possibilities the chemist can describe?

YOU WILL NEED TO KNOW

Concepts

1. A general notion as to the meaning of energy and the means by which it may be transferred from one object to another.

2. How to add two or more chemical equations to give a single equation. In this regard, chemical equations are treated as though they were algebraic equations. See the discussion of Hess's Law in the text — Section 6.2.

Math

1. How to work problems in stoichiometry. See Skills 1 to 4 of Chapter 4 in this guide.

BASIC SKILLS

1. Use a thermochemical equation to relate heat flow in a reaction to amounts (moles, grams) of products or reactants.

This skill is illustrated in Example 6.1 and in the following example.

— —

For the reaction $CH_4(g) + 2\ O_2(g) \rightarrow CO_2(g) + 2\ H_2O(1)$, $\Delta H = -890$ kJ. How many grams of CH_4 must be burned to evolve 1.00 kJ of heat? _____

Here, as always in working problems dealing with balanced equations, we follow a conversion factor approach. The thermochemical equation gives us a conversion factor we need, that between moles of methane and amount of heat evolved.

$$1 \text{ mol } CH_4 = -890 \text{ kJ}$$

$$\text{but also, } = 16.0 \text{ g } CH_4$$

To evolve 1.00 kJ, we need:

$$\text{g } CH_4 = -1.00 \text{ kJ} \times \frac{1 \text{ mol } CH_4}{-890 \text{ kJ}} \times \frac{16.0 \text{ g } CH_4}{\text{mol } CH_4}$$

$$= 0.0180 \text{ g } CH_4$$

— —

See the catalog of problems at the end of the chapter for further practice in using this skill.

2. Relate the enthalpy changes for two reactions whose equations differ only in the direction of reaction and/or in the values of the coefficients used.

This straightforward skill is illustrated in the text by Example 6.2. The skill applies two principles stated in Section 6.2:

(a) ΔH is directly proportional to the amount of substance that reacts or is produced in a reaction. (So, for example, if all coefficients in an equation are multiplied by the number n, then ΔH for the reaction is also multiplied by n.)

(b) ΔH for a reaction is equal in magnitude but opposite in sign to ΔH for the reverse reaction.

Further practice is given in the problems at the end of the chapter and in the Self-Test in this chapter.

3. Calculate ΔH for a reaction from heats of formation.

This is readily accomplished by applying Equation 6.8 to the data of Tables 6.1 and 6.2 Be careful about signs. Since most heats of formation are negative, the products will ordinarily make a negative contribution to ΔH, the reactants a positive contribution. Examples 6.3 and 6.4 illustrate this skill in its simplest form. For additional practice, see the catalog of problems.

4. Calculate ΔH for a reaction from bond energies (enthalpies).

The data come from Table 6.3 in the text. See Example 6.5 for an illustration of this skill. Note that to use the skill, you must:

– know what bonds are present in reactant and product molecules. (The formula HCl must refer to the bonding of a hydrogen atom to a chlorine atom. But what if you were given NH_3? What atoms are bonded?) At this stage, prior to a discussion of chemical bonding (Chapter 10), the nature of the bonds will be indicated in the statement of the problem.

– apply Hess's Law (see the discussion in the text). We make extensive use of this law in this and later chapters.

5. Use calorimetric data to determine the heat flow, Q, for a reaction.

The principle being applied here is that the heat given off by the reaction is absorbed by the calorimeter and its contents. In a coffee-cup calorimeter containing water, this simplifies to

$$Q_{reaction} = -(Q_{water})$$

$$= -(4.18 \; \frac{J}{g \cdot °C} \times m_{water}) \Delta t$$

See Example 6.6 for an application of this skill.

If a bomb calorimeter is used, we must consider both the heat absorbed by the water present and that taken up by the calorimeter container itself. This consideration leads to Equation 6.19 in the text or, going one step further, as in Example 6.7:

$$Q_{reaction} = -(4.18 \ \frac{J}{g \cdot °C} \ \times \ m_{water} + C)\Delta t$$

where C is the calorimeter constant (J/°C). See the catalog of problems for additional illustrations of the skill.

6. (Optional.) Use the First Law of Thermodynamics to calculate ΔE, W, or Q, given the other two quantities. Relate ΔH and ΔE for a reaction.

A simple calculation of each type is discussed in the body of the text, Section 6.6. If you don't like the minus sign in Equation 6.26, blame it on the engineers who decided a long time ago to regard Q as positive when heat flows into a system, but took W to be negative when work "flows into" (is done upon) a system.

Two notes concerning the second part of the skill, one practical, the other philosophical:

− ΔH is the quantity measured when a reaction is carried out in an open container. It is also the quantity calculated when Tables 6.1 to 6.3 are used. ΔE is the quantity measured when a reaction is carried out in a sealed container such as a bomb calorimeter. The difference between ΔH and ΔE is always small, but thermodynamicists tend to be perfectionists who worry about little things.

− Equations 6.26 and 6.29 are typical of thermodynamics: very simple equations based on precise and somewhat subtle lines of reasoning. The equations are easy to use. The only difficulty is to decide under what conditions they are applicable.

SELF-TEST

True or False

1. In carrying out a reaction in a test tube, a student observes (   ) that the test tube becomes cold. He should call the reaction exothermic.

2. Given the thermochemical equation   (   )

$$UF_6(1) \rightarrow UF_6(g); \Delta H = +30.1 \text{ kJ},$$

you can be sure that 30.1 kJ of heat must be evolved whenever 1 mol of liquid UF_6 evaporates.

3. The enthalpy of formation is defined as zero for both   (   )
$H^+(aq)$ and any element in its stable form at 25°C and 1 atm.

4. You would expect that the heat produced in the following   (   )
reactions would have one and the same numerical value:

$$H_2(g) + \frac{1}{2} O_2(g) \rightarrow H_2O(1); \quad 2 H_2(g) + O_2(g) \rightarrow 2 H_2O(1)$$

5. The difference in enthalpy between 2 mol Cl and 1 mol Cl_2,   (   )
both at 25°C and 1 atm, is equal in magnitude to the B.E. of Cl_2.

6. You would expect that ΔH for the following reaction would   (   )
be less negative than the molar heat of formation of $H_2O(1)$:

$$2 H(g) + O(g) \rightarrow H_2O(1)$$

Multiple Choice

7. In which one of the following systems is the amount of   (   )
stored energy the largest?
    (a) 1 mol $H_2O(1)$ at 0°C, 1 atm
    (b) 1 mol $H_2O(1)$ at 100°C, 1 atm
    (c) 1 mol $H_2O(g)$ at 100°C, 1 atm
    (d) b and c are the same

8. The molar heat of combustion of methane, CH_4, is reported   (   )
as –890 kJ. The corresponding thermochemical equation is
    (a) $C(g) + 4 H(g) \rightarrow CH_4(g)$; $\Delta H = -890 \text{ kJ}$
    (b) $C(s) + 2 H_2(g) \rightarrow CH_4(g)$; $\Delta H = -890 \text{ kJ}$
    (c) $CH_4(g) + 3/2.O_2(g) \rightarrow CO(g) + 2 H_2O(1)$; $\Delta H = -890 \text{ kJ}$
    (d) $CH_4(g) + 2 O_2(g) \rightarrow CO_2(g) + 2 H_2O(1)$; $\Delta H = -890 \text{ kJ}$

9. When water evaporates at constant pressure, the *sign* of the   (   )
heat flow
    (a) is negative   (b) is positive
    (c) depends on   (d) depends on container
         temperature   volume

10. Given $\Delta H = -601.8 \text{ kJ}$ for the reaction   (   )

$$Mg(s) + \frac{1}{2} O_2(g) \rightarrow MgO(s),$$

what would you expect to happen if the reaction were allowed to proceed at 1 atm in such a way that no heat flow could occur between the reaction mixture and the surroundings?

(a)   no reaction could occur
(b)   the temperature of the reaction mixture would rise
(c)   the temperature of the reaction mixture would drop
(d)   not enough information is given

11.   When 1.00 g NH_3 (molecular mass = 17.0) is formed from (   )
the elements at 25°C and 1 atm, 2720 J are evolved. The molar heat of formation of NH_3, in kilojoules, is

(a)   −2.72(17.0)   (b)   17.0/2720
(c)   −2.72/17.0   (d)   +2.72(17.0)

12.   Given that ΔH = +185 kJ for the reaction   (   )

$$2\ HCl(g) \rightarrow H_2(g) + Cl_2(g),$$

the H−Cl bond energy

(a)   is −185 kJ   (b)   is −92.5 kJ
(c)   is +92.5 kJ   (d)   cannot be known
                                          without more
                                          information

13.   The bond energies of the halogens F_2, Cl_2, Br_2, and I_2 are (   )
153 kJ, 243 kJ, 193 kJ, and 151 kJ, respectively. Of these, the strongest bond is found in

(a) F_2 (b) Cl_2
(c) Br_2 (d) I_2

14.   The sign of ΔH for the reaction $X_2(g) \rightarrow 2\ X(g)$, where X is (   )
an atom

(a)   is negative   (b)   is positive
(c) $\Delta H = 0$ (d)   depends on the
                                          identity of X

15.   Consider the reaction: $2\ H(g) + O(g) \rightarrow H_2O(1)$. For this (   )
reaction, ΔH is the same as

(a) ΔH_f $H_2O(1)$ (b) ΔH_f $H_2O(g)$
(c)   −2 × B.E. (O−H)   (d)   none of these

16.   Which property of a mole of a pure substance depends on (   )
the pressure and temperature?

(a)   H   (b)   E
(c)   V   (d)   all of these

17. In order to determine the heat evolved when a sample is ( )
burned in a bomb calorimeter, you must know
     (a) the mass and specific heat of the H_2O in the calo-
        rimeter
     (b) the temperature change
     (c) the calorimeter constant
     (d) all the above

18. If 20 g $H_2O(1)$ which originally were at 25°C are heated to ( )
35°C, the heat absorbed by the water is about
     (a) +200 J   (b) +600 J
     (c) +800 J   (d) -800 J

19. Current sources of energy include coal, oil and natural gas, ( )
water power, and nuclear reactions. Which one of these sources
seems most likely to decrease in importance over the next fifty
years?
     (a) coal   (b) oil and natural gas
     (c) water power   (d) nuclear reactions

20. The greatest potential for meeting our energy needs in the ( )
long run seems to be offered by
     (a) wood   (b) petroleum
     (c) fission   (d) fusion on the earth or
                                   sun

Problems

1. Using the data below, calculate ΔH for each reaction at
25°C, 1 atm.
     (a) $2 \ KClO_3(s) \rightarrow 2 \ KCl(s) + 3 \ O_2(g)$
     (b) $Cl_2(g) + 2 \ Br^-(aq) \rightarrow 2 \ Cl^-(aq) + Br_2(1)$

	ΔH_f (kJ/mol)		
$Br^-(aq)$	-121	$KCl(s)$	-436
$Cl^-(aq)$	-167	$KClO_3(s)$	-391

2. Using your result for Problem 1(a), determine the heat flow for the
decomposition of 1.72 g $KClO_3$. (atomic masses: O = 16.0, Cl = 35.5, K =
39.1)

3. For the reaction, $2 \ C_2H_2(g) + 5 \ O_2(g) \rightarrow 4 \ CO_2(g) + 2 \ H_2O(1)$, ΔH
= -2599.0 kJ. The molar heats of formation of $CO_2(g)$ and $H_2O(1)$ are
-393.5 kJ and -285.8 kJ, respectively. Calculate:
     (a) the molar heat of formation of acetylene, C_2H_2

(b) the final temperature reached if all the heat evolved in the reaction of 2 mol C_2H_2 is absorbed by 50.0 kg H_2O, originally at 20.0°C (specific heat H_2O = 4.18 J/(g·°C))

4. From the data at 298 K, 1 atm: $\Delta H(kJ)$

½ $H_2(g)$ + ½ $O_2(g)$ → OH(g)   +42.09

$H_2(g)$ + ½ $O_2(g)$ → $H_2O(g)$ −241.80

$H_2(g)$ → 2 H(g)   +435.89

calculate ΔH for the reaction H(g) + OH(g) → $H_2O(g)$.

5. When 1.80 g of steam at 100.00°C is allowed to condense in a bomb calorimeter originally at 25.00°C, the final temperature is 30.00°C. What is the calorimeter constant in joules per degree Celsius? (ΔH_{vap} H_2O = 40.63 kJ/mol; specific heat H_2O = 4.18 J/(g·°C); AM H = 1.0, O = 16.0)

SELF-TEST ANSWERS

1. **F** (To restore the temperature to its initial value, heat must be absorbed by the test tube.)
2. **F** (Absorbed. Be careful with the *sign* of thermochemical quantities.)
3. **T**
4. **F** (The second equation is taken to mean twice as many moles, so the energy change is twice as great.)
5. **T** (All species would have to be gaseous.)
6. **F** (The given reaction can be considered as the sum of the two reactions: $H_2(g)$ + ½ $O_2(g)$ → $H_2O(1)$ and 2 H(g) + O(g) → $H_2(g)$ + ½ $O_2(g)$. The second step releases additional energy, making the overall ΔH more negative.)
7. **c** (Evaporation absorbs heat, so more energy is stored in the vapor.)
8. **d** (Note that CO_2 is the expected product of combustion.)
9. **b** (Choice c is ruled out since ΔH doesn't change much with T.)
10. **b**
11. **a** (Check the units as well as the sign.)
12. **d** (To find: need ΔH for the reaction HCl(g) → H(g) + Cl(g).)
13. **b** (The stronger bond is the more difficult to break.)
14. **b** (Energy must be absorbed to break any bond.)
15. **d** (Not c, because the product is not gaseous.)
16. **d** (From another point of view: specific values of pressure and temperature will automatically determine the values of H, E, V, and other state properties.)
17. **d**

18. c (Approximately, $Q = 4J/(g \cdot °C) \times 20 \text{ g} \times 10°C = 800 \text{ J}$.)
19. b
20. d (That is, fusion reactors or solar energy.)

Solutions to Problems

1. (a) $\Delta H = 2 \Delta H_f \text{ KCl} - 2 \Delta H_f \text{ KClO}_3 = -90 \text{ kJ}$
   (b) $\Delta H = 2 \Delta H_f \text{ Cl}^- - 2 \Delta H_f \text{ Br}^- = -92 \text{ kJ}$

2. $\Delta H = 1.72 \text{ g KClO}_3 \times \dfrac{-90 \text{ kJ}}{2 \text{ mol KClO}_3} \times \dfrac{1 \text{ mol KClO}_3}{123 \text{ g KClO}_3} = -0.63 \text{ kJ}$

3. (a) For the reaction as written:
   $\Delta H = -2599.0 = 4 \Delta H_f \text{ CO}_2 + 2 \Delta H_f \text{ H}_2\text{O} - 2 \Delta H_f \text{ C}_2\text{H}_2$
   $\Delta H_f \text{ C}_2\text{H}_2 = +226.7 \text{ kJ/mol}$
   (b) $Q_{water} = +2.60 \times 10^6 \text{ J} = 4.18 \text{ J}/(g \cdot °C) \times 5.00 \times 10^4 \text{ g} \times \Delta t$
   $\Delta t = 12.4°C$
   $t_{final} = 32.4°C$

4. The sum of reactions we want is:
$H \rightarrow \frac{1}{2} H_2$	$\Delta H_1 = -217.95 \text{ kJ}$
$OH \rightarrow \frac{1}{2} H_2 + \frac{1}{2} O_2$	$\Delta H_2 = -42.09 \text{ kJ}$
$H_2 + \frac{1}{2} O_2 \rightarrow H_2O$	$\Delta H_3 = -241.80 \text{ kJ}$

   $\Delta H = \Delta H_1 + \Delta H_2 + \Delta H_3 = -501.84 \text{ kJ}$

5. $Q_{steam} = 1.80 \text{ g} \times \dfrac{-40630 \text{ J}}{1.80 \text{ g}} + 1.80 \text{ g} \times -70.00°C \times 4.18 \text{ J}/(g \cdot °C)$
   $= -4060 \text{ J} - 527 \text{ J} = -4590 \text{ J}$
   $Q_{bomb} = 4590 \text{ J} = 5.00°C \times C; C = 918 \text{ J}/°C$

SELECTED READINGS

Energy sources are considered in:

Bamberger, C.E., Hydrogen: A Versatile Element, *American Scientist* (July-August 1975), pp. 438–447.

Bethe, H.A., The Necessity of Fission Power, *Scientific American* (January 1976), pp. 21–31.

Cheney, E.S., U.S. Energy Resources: Limits and Future Outlook, *American Scientist* (January-February 1974), pp. 14–22.

Cromie, W.J., Which Is Riskier — Windmills or Reactors? *SciQuest* (March 1980), pp. 6–10.

Daniels, F., *Direct Use of the Sun's Energy,* New Haven, Yale, 1964.

Energy and Power, *Scientific American* (September 1971).

Fickett, A.P., Fuel-Cell Power Plants, *Scientific American* (December 1978), pp. 70–76.

Holdren, J., *Energy,* New York, Sierra Club, 1971.

Johnston, W.D., The Prospects for Photovoltaic Conversion, *American Scientist* (November-December 1977), pp. 729–736.

Kulcinski, G.L., Energy for the Long Run: Fission or Fusion? *American Scientist* (January-February 1979), pp. 78–89.

Poole, A.D., Flower Power: Prospects for Photosynthetic Energy, *Bulletin of the Atomic Scientists* (May 1976), pp. 49–58.

Walters, E.A., An Overview of the Energy Crisis, *Journal of Chemical Education* (May 1975), pp. 282–288.

Thermodynamics is introduced in three completely different ways in:

Faraday, M., *The Chemical History of a Candle,* New York, Viking, 1960.

Mahan, B.H., *Elementary Chemical Thermodynamics,* New York, W.A. Benjamin, 1963.

Pimentel, G.C., *Understanding Chemical Thermodynamics,* San Francisco, Holden-Day, 1969.

7

PHYSICAL BEHAVIOR OF GASES

QUESTIONS TO GUIDE YOUR STUDY

1. What materials can you think of that usually exist as gases? What properties do they have in common?

2. What conditions favor the existence of a substance as a gas rather than as a liquid or a solid?

3. How do you measure properties of gases such as temperature and pressure? How would you determine the mass of a sample of gas?

4. Is there a simple relationship among the properties of a gas that generally holds for all gases? Can you rationalize such a relationship in terms of atomic-molecular theory? (For example: how does the behavior of molecules explain the relation between temperature and pressure for the air inside a tire?)

5. How do mixtures of gases behave? How is their behavior related to that of a pure gaseous substance?

6. Can the quantities of gases participating in a chemical reaction be simply expressed in terms of masses, moles, or volumes?

7. How does the volume of gaseous reactant or product depend on reaction conditions?

8. What are some of the practical applications (as well as support for other chemical principles) of our knowledge of gas behavior?

9. What experimental support is there for our ideas about the nature of the molecules in a gas?

10. How do you account for the observed differences in properties among gases? Between gases and liquids or gases and solids?

YOU WILL NEED TO KNOW

This chapter is primarily quantitative in nature. It uses the concepts of molecular mass, mole, and Avogadro's number N (Chapter 2) in numerous calculations.

Math

1. How to solve first and second order equations for any one variable. Examples: Rewrite PV = nRT in the form n = ... Or, again, solve the equation ½ mu² = cT for u.

2. How to find the square root of a number. This is most simply done using a calculator. Note that square roots can be eliminated from an equation by squaring both sides. See Skill 6 of this chapter.

3. How to work problems in stoichiometry. See Skills 1 to 4 of Chapter 4 in this guide.

BASIC SKILLS

You will find that material in this chapter can best be mastered by working lots of problems. The units and indeed the equations used in calculations will depend to some extent upon whether you are using the regular version of *Chemical Principles* or the SI version. Throughout this section and the self-test that follows, *calculations which are appropriate for the SI version are set apart in boxes outlined by solid lines.*

1. **Relate the various units of pressure, volume, and temperature to one another.**

This is really not a new skill. It is the use of conversion factors to change units (Chapter 1) all over again. In its simplest form, as in Example 7.1, you are asked to convert between pressure units. In other cases, you may be required to convert volume units. Appropriate conversion factors, taken from Chapter 1 of the text, are:

Pressure: 1 atm = 101.3 kPa = 760 mm Hg

Volume: 1 ℓ = 1 dm³ = 10^3 cm³ = 10^{-3} m³

A sample of gas at a certain temperature and pressure has a volume of 4.29 ℓ. Express this volume in

(a) cubic decimetres_____ (b) cubic centimetres_____ (c) cubic metres_____

Using the conversion factors listed above, you should find that:

V = 4.29 dm³ = 4.29 × 10^3 cm³ = 4.29 × 10^{-3} m³

Note that the volume in cubic decimetres is the same as that in litres. Indeed, one can think of a litre as the volume occupied by a cube 1 dm (10 cm) on an edge.

_ _

Frequently, you will find it necessary to convert from degrees Celsius to kelvin:

$$K = °C + 273$$

In all calculations involving gases, temperature must be expressed in K.

2. Use the Ideal Gas Law to:

a.   Determine the effect of a change in conditions (e.g., a change in T or P) upon a particular variable (e.g., V).

_ _

A sample of a gas has a volume of 312 cm^3 at 273 K and 750 mm Hg. What volume will the sample occupy at 298 K and 0.500 atm?_____

For the gas in both states, PV = nRT. But, from the statement of the problem, the number of moles of gas remains constant. So, for each state we can say PV = nRT = (constant)T. Or, distinguishing states 1 and 2,

$$\frac{P_1 V_1}{T_1} = \frac{P_2 V_2}{T_2} = nR = constant$$

Solving for the variable we need, V_2

$$V_2 = \frac{P_1 V_1}{T_1}\left(\frac{T_2}{P_2}\right)$$

Or, showing the effects of temperature and pressure on volume (e.g., V is directly proportional to T) a little more clearly:

$$V_2 = V_1 \times \frac{T_2}{T_1} \times \frac{P_1}{P_2}$$

In using this equation:

— temperatures must be expressed in K.

— the pressures must be expressed in the same units, either millimetres of mercury or atmospheres. We might convert P_1 to atmospheres:

$$P_1 = 750 \text{ mm Hg} \times \frac{1 \text{ atm}}{760 \text{ mm Hg}} = 0.987 \text{ atm}$$

Now we are ready to substitute into the equation:

$$V_2 = 312 \text{ cm}^3 \times \frac{298 \text{ K}}{273 \text{ K}} \times \frac{0.987 \text{ atm}}{0.500 \text{ atm}} = 672 \text{ cm}^3$$

A similar calculation, using SI units, is shown below.

A sample of a gas has a volume of 1.62 dm^3 at 308 K and 0.915 atm. What volume will the sample occupy at 323 K and 94.3 kPa? _____
As in the preceding example, we use the equation:

$$V_2 = V_1 \times \frac{T_2}{T_1} \times \frac{P_1}{P_2}$$

The two pressures must be expressed in the same units. We convert P_1 to kilopascals:

$$P_1 = 0.915 \text{ atm} \times \frac{101.3 \text{ kPa}}{1 \text{ atm}} = 92.7 \text{ kPa}$$

Now, substituting in the equation for V_2:

$$V_2 = 1.62 \text{ dm}^3 \times \frac{323 \text{ K}}{308 \text{ K}} \times \frac{92.7 \text{ kPa}}{94.3 \text{ kPa}} = 1.67 \text{ dm}^3$$

Examples 7.2 and 7.3 illustrate this skill. Note that in each of these problems we have derived the relationship needed from the Ideal Gas Law by using simple algebra. In case you are tempted to try to memorize these relationships, we should point out that there are ten of them! Again in each problem solution note that units have been made consistent *and all temperatures are expressed in K.*

   b. **Solve for one variable (e.g., V), given the values of the others** (e.g., P, n, and T).

What volume is occupied by 2.10 mol of an ideal gas at 20°C and 1.50 atm? _____
Solving the Ideal Gas Law for V: V = nRT/P, substituting T in K, and using the appropriate value of R, 0.0821 ℓ·atm/(mol·K):

$$V = \frac{2.10 \text{ mol} \times 0.0821 \dfrac{\ell \cdot \text{atm}}{\text{mol} \cdot \text{K}} \times 293 \text{ K}}{1.50 \text{ atm}} = 33.6 \text{ ℓ}$$

What volume is occupied by 2.10 mol of an ideal gas at 20°C and 152 kPa? _____

Solving the Ideal Gas Law for V: V = nRT/P, substituting T in K, and using the appropriate value of R, 8.31 kPa·dm³/(mol·K):

$$V = \frac{2.10 \text{ mol} \times 8.31 \frac{kPa \cdot dm^3}{mol \cdot K} \times 293 \text{ K}}{152 \text{ kPa}} = 33.6 \text{ dm}^3$$

- -

Example 7.4 is similar, asking you to solve the gas law expression for P instead of V, and requiring that a mass in grams first be converted to a number of moles.

This skill can be combined with the principles regarding mass and mole relationships in chemical reactions (Chapter 4). You can relate the number of moles or mass of a reactant or product to the volume or pressure of a gaseous substance taking part in the reaction. This application is demonstrated in Example 7.7. See the catalog of problems for further illustration of the skill in this form.

c. Calculate the molecular mass of a gas, knowing the mass of a given volume (or the density) at a known P and T.

See Example 7.5 for a straightforward application of this skill.

d. Calculate the density of a gas at a given temperature and pressure.

Example 7.6 shows how the Ideal Gas Law can be applied in this way.

3. Relate the volumes of gases, measured at the same T and P, involved in a chemical reaction.

The basic principle here is: the coefficients of a balanced equation relate not only the numbers of moles of reactants and products, but also the volumes of gaseous reactants and products. Those volumes must, however, be measured at the same temperature and pressure for this to apply. So, for the reaction: $2 H_2(g) + O_2(g) \rightarrow 2 H_2O(g)$, if we started with 50 cm³ of H_2, 25 cm³ of O_2 would be required to react with it and 50 cm³ of water vapor would be produced, all at the same T and P. See Example 7.8 for another illustration of this skill.

Note that, for the formation of water cited above, if the product had been $H_2O(1)$, its volume would have been a great deal less than 50 cm^3. Or again, if the gases had been measured at different pressures or temperatures, the 2:1:2 volume relationship would not hold.

4. Use Dalton's Law to obtain partial pressures of gases in mixtures.

This is shown in Example 7.9. There, as is the usual case in general chemistry, the law is used to obtain the partial pressure of a gas which is collected over water. This is often the first step in a calculation involving measurements of gases in the laboratory.

5. Relate the mole fraction of a gas to its partial pressure.

Example 7.10 gives a simple illustration of this skill. The mole fraction (X) of gas A in a mixture with total number of moles of gas n_{tot} is defined by the relation: $X_A = n_A/n_{tot}$, where n_A is the number of moles of gas A in the mixture. The mole fraction is related to the partial pressure: $P_A = X_A P_{tot}$, where P_{tot} is the directly measurable total pressure of the mixture and is the sum of all the partial pressures.

6. Use Graham's Law to relate the molecular masses of two gases to their rates, or times, of effusion.

- -

It is found that the rate of effusion of a certain gas is 0.600 times that of N_2 (MM = 28.0) under the same conditions. What is the molecular mass of the gas? _____

Applying Graham's Law, we have:

$$\frac{rate_x}{rate_{N_2}} = \left(\frac{MM_{N_2}}{MM_x}\right)^{1/2}$$

but since $rate_x = 0.600\ rate_{N_2}$, we can write:

$$\frac{rate_x}{rate_{N_2}} = 0.600$$

and substitute in the original expression:

$$0.600 = \left(\frac{28.0}{MM_x}\right)^{1/2}$$

Squaring both sides and solving for MM_x:

$$0.360 = \frac{28.0}{MM_x} \, , \; MM_x = \frac{28.0}{0.360} = 77.8$$

– –

See Example 7.11, where times of effusion, rather than rates, are given in the statement of the problem. The time for an event to take place is inversely proportional to the rate at which it occurs. Hence, the time of effusion turns out to be *directly* proportional to the square root of molecular mass.

7. **Calculate the average speed of molecules of a particular gas at a given temperature.**

Example 7.12 illustrates the use of this skill. Note that
– to calculate the average speed in centimetres per second, you must use the molar mass in grams and R = 8.31×10^7 g·cm^2/(s^2·mol·K). Thus, to obtain the average velocity of an H_2 molecule (GMM = 2.016 g/mol) at 25°C:

$$u \text{ (in cm/s)} = \left(\frac{3RT}{GMM}\right)^{1/2} = \left(\frac{3 \times 8.31 \times 10^7 \times 298}{2.016}\right)^{1/2} = 1.92 \times 10^5 \text{ cm/s}$$

– to calculate the average velocity in metres per second, you should use R = 8.31 kg·m^2/(s^2·mol·K) *and express the molar mass in kilograms.* Thus, to obtain the average velocity of an H_2 molecule (KMM = 2.016×10^{-3} kg/mol) at 25°C:

$$u \text{ (in m/s)} = \left(\frac{3RT}{KMM}\right)^{1/2} = \left(\frac{3 \times 8.31 \times 298}{2.016 \times 10^{-3}}\right)^{1/2} = 1.92 \times 10^3 \text{ m/s}$$

SELF-TEST

True or False

1. At 25°C and 1 atm, the molecules in a sample of air are ( ) separated by distances that are large compared to molecular diameters.

2. The volume of a sample of gas is directly proportional to its   (   )
Kelvin temperature.

3. One atmosphere is about the same as 100 kPa.   (   )

4. The average kinetic energies of O_2 and H_2 molecules, both (   )
gases at the same temperature, are in the same ratio as their
molecular masses.

5. The average speed of a gas molecule depends only on the   (   )
absolute temperature.

6. The densities of two samples of air at 1 atm compare as do   (   )
their Kelvin temperatures.

7. In the process known as effusion, two or more gases mix to   (   )
form a solution.

Multiple Choice

8. Which of the following substances do you expect to be   (   )
gaseous at room temperature and atmospheric pressure?
  (a) CO_2 (b) NH_3
  (c) HCl   (d) all of these

9. The volume of a mole of gas   (   )
  (a) is 22.4 ℓ
  (b) depends on the composition of the gas
  (c) is directly proportional to T, inversely to P
  (d) is directly proportional to P, inversely to T

10. A certain mountain rises 3.5 km above sea level. The   (   )
pressure at the top is about 2/3 that at sea level, and the Kelvin
temperature is 9/10 as great. If a balloon is inflated at sea level and
then carried to the mountain top, by what factor does its volume
change?
  (a) 18/30   (b) 27/20
  (c) 20/27   (d) 30/18

11. When two gases are mixed, the pressure of the mixture is   (   )
  (a) given by Gay-Lussac's Law of Combining Volumes
  (b) the product of the pressures each would have if alone
  (c) the difference of the pressures each would have if alone
  (d) the sum of the pressures each would have if alone

12. At the same T and P, two flasks of equal volume are filled   (   )
with different gases, A and B. The mass of A is 0.34 g, while that of
B is 0.48 g. It is known that B is ozone, O_3 (molecular mass = 48),

and that A is one of the following. Which one is most likely to be gas A?
- (a) O_2 (molecular mass = 32)
- (b) H_2S (molecular mass = 34)
- (c) SO_2 (molecular mass = 64)
- (d) any one of these is possible

13. Consider the gases of question 12 again. Which statement ( ) below is true?
- (a) the numbers of molecules of A and B are equal
- (b) the masses of individual molecules of A and B compare in the same way as the masses of the samples
- (c) the average translational energies of A and B molecules are the same
- (d) all of these statements are true

14. The Ideal Gas Law only approximately describes the ( ) behavior of a real gas. This can be partly explained by the idea that
- (a) R is not really a constant
- (b) gas molecules have zero volume
- (c) gas molecules interact with each other
- (d) translational energy is not really directly proportional to T

15. Real gases behave most nearly like the Ideal Gas Law says ( ) they do at
- (a) high T, low P
- (b) low T, high P
- (c) high T, high P
- (d) low T, low P

16. Your lecturer opens a bottle of H_2S and a bottle of ether, ( ) $C_4H_{10}O$, at the same time. Both are gases at the same temperature and pressure. Which smelly gas should you detect first?
- (a) H_2S (molecular mass = 34)
- (b) $C_4H_{10}O$ (molecular mass = 74)
- (c) both at the same time
- (d) neither, since they are more dense than air

17. What must be the molecular mass of a gas that effuses ( ) one-fourth as rapidly as CH_4 (molecular mass = 16)?
- (a) 4
- (b) 16
- (c) 64
- (d) 256

18. When 20 cm^3 of N_2 and 60 cm^3 of H_2 are mixed and ( ) allowed to react to form $NH_3(g)$, what volume of NH_3 is formed at the same T and P? (Assume 100% yield.)
- (a) 20 cm^3
- (b) 40 cm^3
- (c) 60 cm^3
- (d) 120 cm^3

19. In a mixture of H_2, He and Ne at 25°C, the molecules with   ( )
the greatest average speed are
    (a) H_2 (molecular mass = 2)
    (b) He (molecular mass = 4)
    (c) Ne (molecular mass = 20)
    (d) all speeds are the same

20. The relation between n and T (constant P, V) is: ( )

    (a) $\dfrac{n_2}{n_1} = \dfrac{T_2}{T_1}$ (b) $\dfrac{n_2}{n_1} = \dfrac{T_1}{T_2}$

    (c) $\dfrac{n_2}{n_1} = \left(\dfrac{T_2}{T_1}\right)^{1/2}$ (d) $\dfrac{n_2}{n_1} = \left(\dfrac{T_1}{T_2}\right)^{1/2}$

Problems

(If you are using the SI version of *Chemical Principles,* work the boxed problems on page 71).

1. The pressure of a gas sample is 2.64 atm when it occupies a 250 cm³ flask at 25°C. What will its pressure become if the gas is transferred to a 500 cm³ flask at 0°C?

2. An N_2 tank having a volume of 5.67 ℓ weighs 8.43 kg when full and 6.93 kg when empty. Calculate the pressure in atmospheres of the N_2 (molecular mass = 28.0) when the tank is full at 22°C.

3. The density of a certain gaseous compound of nitrogen and oxygen is 2.00 g/ℓ at 297 K and 1.06 atm. What is the molecular mass of the compound? Its molecular formula? (AM N = 14.0, O = 16.0)

4. A 5.00 ℓ flask contains a mixture of an unknown gas and C_3H_6 (molecular mass = 42.0) at 27°C and a total pressure of 1.50 atm. Analysis of the mixture shows that 2.10 g of C_3H_6 are present.
    (a) Calculate the partial pressure of C_3H_6 in atmospheres.
    (b) Calculate the mole fraction of the unknown gas.

5. A sample of PH_3 occupies a volume of 529 cm³ at 77°C and 742 mm Hg. If it is decomposed completely to the elements (P_4 and H_2) at this temperature and pressure, what will be the volume of the gaseous products?

SI Problems

1. The pressure of a gas sample is 267 kPa when it occupies a 250 cm³ flask at 25°C. What will its pressure become if the gas is transferred to a 500 cm³ flask at 0°C?

2. An N_2 tank having a volume of 5.67 dm³ weighs 8.43 kg when full and 6.93 kg when empty. Calculate the pressure in kilopascals of the N_2 (molecular mass = 28.0) when the tank is full at 22°C.

3. The density of a certain gaseous compound of nitrogen and oxygen is 2.00 g/dm³ at 297 K and 107 kPa. What is the molecular mass of the compound? Its molecular formula? (AM N = 14.0, O = 16.0)

4. A 5.00 dm³ flask contains a mixture of an unknown gas and C_3H_6 (molecular mass = 42.0) at 27°C and a total pressure of 152 kPa. Analysis of the mixture shows that 2.10 g of C_3H_6 is present.
    (a)   Calculate the partial pressure of C_3H_6 in kilopascals.
    (b)   Calculate the mole fraction of the unknown gas.

5. A sample of PH_3 occupies a volume of 529 cm³ at 77°C and 98.9 kPa. If it is decomposed completely to the elements (P_4 and H_2) at this temperature and pressure, what will be the volume of the gaseous products?

SELF-TEST ANSWERS

1. **T** (This explains, for example, why gases are so compressible.)
2. **T** (The law of Charles and Gay-Lussac: V = constant × T.)
3. **T** (To be exact, 1 atm = 101.3 kPa.)
4. **F** (At the same T, the average translational energies are the same.)
5. **F** (A subtle question. For a given molecule, this is true. Otherwise, it depends upon molecular mass. See Equation 7.20.)
6. **F** (Density is inversely proportional to T. See Equation 7.8.)
7. **F** (A gas escapes through a small opening into a vacuum or another gas.)
8. **d** (Some of the gaseous substances you should recognize.)
9. **c** (V = nRT/P. Choice a is appropriate only at 273 K, 1 atm.)
10. **b** $(V = \dfrac{3}{2} \times \dfrac{9}{10} \times \dfrac{27}{20} \times$ original volume.)
11. **d**

12. b   (At the same P, V, and T there must be equal numbers of moles or of molecules. So masses in grams are in the same ratio as molecular masses.)

13. d

14. c   (Also, the molecules themselves have volume.)

15. a   (Minimizing the effects of both attractive forces and molecular volume.)

16. a   (At the same T, molecules of smaller mass move faster.)

17. d   (Set up the Graham's Law expression.)

18. b   (The balanced equation for the reaction is $N_2(g) + 3\ H_2(g) \rightarrow 2\ NH_3(g)$.)

19. a   (See Question 16.)

20. b   (From the Ideal Gas Law $PV = nRT$, n is *inversely* proportional to T.)

Solutions to Problems

1. $P_2 = P_1 \times \dfrac{T_2}{T_1} \times \dfrac{V_1}{V_2} = 2.64\ \text{atm} \times \dfrac{273\ K}{298\ K} \times \dfrac{250\ cm^3}{500\ cm^3} = 1.21\ \text{atm}$

2. $P = \dfrac{nRT}{V}$; $n = \dfrac{1.50 \times 10^3\ g}{28.0\ g/mol} = 53.6\ mol$

    $P = \dfrac{53.6\ mol \times 0.0821\ \dfrac{\ell \cdot atm}{mol \cdot K} \times 295\ K}{5.67\ \ell} = 229\ \text{atm}$

3. $GMM = \dfrac{dRT}{P} = \dfrac{2.00\ g/\ell \times 0.0821\ \dfrac{\ell \cdot atm}{mol \cdot K} \times 297\ K}{1.06\ atm} = 46.0\ g/mol$

    formula: NO_2

4. (a) For the C_3H_6: $P = \dfrac{nRT}{V}$; $n = \dfrac{2.10\ g}{42.0\ g/mol} = 5.00 \times 10^{-2}\ mol$

    $P = \dfrac{5.00 \times 10^{-2}\ mol \times 0.0821\ \dfrac{\ell \cdot atm}{mol \cdot K} \times 300\ K}{5.00}$

    $= 0.246\ \text{atm}$

    (b) For the unknown gas: $P = 1.50\ \text{atm} - 0.25\ \text{atm} = 1.25\ \text{atm}$

    $X = \dfrac{1.25\ atm}{1.50\ atm} = 0.833$

5. The equation for the reaction is:

    $$4\ PH_3(g) \rightarrow P_4(g) + 6\ H_2(g)$$

The volume of the products must then be 7/4 that of the reactant

$$V = 7/4 \times 529\ cm^3 = 926\ cm^3$$

Solutions to SI Problems

1. $P_2 = P_1 \times \dfrac{T_2}{T_1} \times \dfrac{V_1}{V_2} = 267 \text{ kPa} \times \dfrac{273 \text{ K}}{298 \text{ K}} \times \dfrac{250 \text{ cm}^3}{500 \text{ cm}^3} = 122 \text{ kPa}$

2. $P = \dfrac{nRT}{V}$; $n = \dfrac{1.50 \times 10^3 \text{ g}}{28.0 \text{ g/mol}} = 53.6 \text{ mol}$

$$= \dfrac{53.6 \text{ mol} \times 8.31 \dfrac{\text{kPa} \cdot \text{dm}^3}{\text{mol} \cdot \text{K}} \times 295 \text{ K}}{5.67 \text{ dm}^3} = 2.32 \times 10^4 \text{ kPa}$$

3. $\text{GMM} = \dfrac{dRT}{P} = \dfrac{2.00 \text{ g/dm}^3 \times 8.31 \dfrac{\text{kPa} \cdot \text{dm}^3}{\text{mol} \cdot \text{K}} \times 297 \text{ K}}{107 \text{ kPa}} = 46.0 \text{ g/mol}$

formula: NO_2

4. (a) For the C_3H_6: $P = \dfrac{nRT}{V}$; $n = \dfrac{2.10 \text{ g}}{42.0 \text{ g/mol}} = 5.00 \times 10^{-2} \text{ mol}$

$$P = \dfrac{5.00 \times 10^{-2} \text{ mol} \times 8.31 \dfrac{\text{kPa} \cdot \text{dm}^3}{\text{mol} \cdot \text{K}} \times 300 \text{ K}}{5.00 \text{ dm}^3}$$

$$= 24.9 \text{ kPa}$$

(b) For the unknown gas: $P = 152 \text{ kPa} - 25 \text{ kPa} = 127 \text{ kPa}$

$X = \dfrac{127 \text{ kPa}}{152 \text{ kPa}} = 0.836$

5. The equation for the reaction is:

$$4 \text{ PH}_3 (g) \rightarrow P_4 (g) + 6 \text{ H}_2 (g)$$

The volume of the products must then be 7/4 that of the reactant:

$$V = \text{7/4} \times 529 \text{ cm}^3 = 926 \text{ cm}^3$$

SELECTED READINGS

Gases at low pressures and temperatures, and a modern application of gas laws:

Cohen, E.G.D., Toward Absolute Zero, *American Scientist* (November-December 1977), pp. 752–758.

Olander, D.R., The Gas Centrifuge, *Scientific American* (August 1978), pp. 37–43.

Proctor, W.G., Negative Absolute Temperatures, *Scientific American* (August 1978), pp. 90–99.

Steinherz, H.A., Ultrahigh Vacuum, *Scientific American* (March 1962), pp. 78–90.

History of investigations into gas behavior is considered in:

Conant, J.B., *Science and Common Sense,* New Haven, Yale, 1951.
Neville, R.G., The Discovery of Boyle's Law, 1661–62, *Journal of Chemical Education* (July 1972), pp. 356–359.

Kinetic theory is extensively discussed, with lots on real behavior, in:

Hildebrand, J.H., *An Introduction to Modern Kinetic Theory,* New York, Reinhold, 1963.
Kauzmann, W., *Kinetic Theory of Gases,* Menlo Park, Calif., W.A. Benjamin, 1966.

8

THE ELECTRONIC STRUCTURE OF ATOMS

QUESTIONS TO GUIDE YOUR STUDY

1. What evidence is there for the idea that atoms are themselves composed of smaller parts? What are the dimensions of a typical atom and of its parts?

2. For an electron inside an atom — where is it, what is it doing?

3. How many arrangements are possible for the electrons in any given atom? Which one is observed under ordinary conditions?

4. How are electron arrangements in atoms represented?

5. What experimental observations support the details of our model of electronic structure?

6. What energy changes are associated with changes in electronic structure? How much energy is required to remove an electron from a neutral atom?

7. How do electronic and thermochemical energies compare? How are the energy changes in chemical reactions related to energy changes of electrons in atoms and molecules?

8. What determines the number of electrons a neutral atom will possess? The number of electrons the atom may lose or gain to form an ion?

9. What properties of isolated atoms can be explained by electron arrangements? What correlations can be made with position in the Periodic Table?

10. Does modern quantum theory provide all the answers?

YOU WILL NEED TO KNOW

Concepts

1. General ideas of atomic and nuclear composition (kinds, numbers, and charges of nuclear particles; meaning of atomic number). Review Chapter 2.

2. A general idea of what composes the electromagnetic spectrum (infrared, visible, ultraviolet ...) — the names of the regions and their

approximate wavelengths. See the current chapter, as well as any introductory physics textbook.

Math

No new math is introduced in the few calculations of this chapter.

BASIC SKILLS

   1. **Relate the wavelength of a spectral line to the change in energy of an atom.**

What this skill involves is the application of Equation 8.1 or, in a more useful form, Equation 8.2. The calculation is straightforward and is illustrated by Example 8.1 in the text. In any such calculation be sure your units are consistent.

   2. **Use the Bohr theory to calculate the energy of an electron in a given principal energy level of the hydrogen atom, or the difference in energy between two levels.**

This may sound more complicated than it really is. It involves the use of Equation 8.4 and is again illustrated by Example 8.1. As noted in this example, it is probably least confusing to first calculate the energies of the individual states, then subtract to find the energy change. See the catalog of problems for additional practice.

The sign of E and of ΔE often causes some concern. For an electron in an atom, the energy is always a negative quantity. This is the result of the way the energy of the electron has been defined (zero at infinite separation from the nucleus). When an electron moves from a low level to a higher one, the sign of the energy change, ΔE, is positive (energy is absorbed by the atom in such a transition). The sign of ΔE is negative when an electron drops from a higher level to a lower level (energy is evolved to the surroundings).

   3. **Determine the number of electrons that may be accommodated by any given principal energy level or sublevel.**

In any principal energy level there are n sublevels. The capacity of a sublevel is given by: s = 2, p = 6, d = 10, f = 14, . . . The total number of electrons which may fit into a principal energy level n is given by $2n^2$.

Example 8.2 illustrates the use of the quantum number rules summarized in this way. What is generally more useful in describing electron arrangements, however, is the next skill.

4. Given the atomic number of an element, write the electron configuration of its isolated gaseous atom in the ground state.

The electron configuration of an atom gives the number of electrons in each sublevel. In order to write the configuration, you must know:

(a)   the capacity of each sublevel

(b)   the order in which the sublevels are filled. Knowing the sequence through 3d (1s, 2s, 2p, 3s, 3p, 4s, 3d) is enough for writing most of the electronic configurations you are likely to need. Almost any other atom of atomic number higher than 30 (at which 3d is filled) can be assigned its configuration by use of the Periodic Table. See Skill 7.

This skill is used in Example 8.3 and serves as the basis for working Example 8.4 as well. Practice in this skill is very useful for much of the chemistry to come. See the catalog of problems at the end of the chapter.

5. Given or having written the electron configuration of an atom, draw its orbital diagram.

To make this conversion, you must realize that:

(a)   each sublevel is divided into orbitals capable of holding two electrons apiece. An s sublevel has one orbital; a p sublevel, three; a d sublevel, five; an f sublevel, seven.

(b)   When there are two electrons in an orbital, they have opposed spins, indicated by ($\uparrow\downarrow$).

(c)   In a partially filled sublevel, there are as many half-filled orbitals as possible. Electrons in these orbitals have the same spin. For example, np^3 would be indicated by ($\uparrow$) ($\uparrow$) ($\uparrow$).

Example 8.4 and the exercise following it require the use of this skill. See the catalog of problems for further practice in this useful skill.

6. Give the four quantum numbers corresponding to each of various electrons in an atom.

You should know the rules for assigning quantum numbers. They may be summarized as follows:

(a) $n = 1, 2, 3, \ldots$

(b) $\ell = 0, 1, 2, \ldots (n-1)$. For an s electron, $\ell = 0$; for p electrons, $\ell = 1$; for d, $\ell = 2$; for f, $\ell = 3$.

(c) m_ℓ can take on any whole-number value, including zero, ranging from $+\ell$ to $-\ell$.

(d) m_s can be either $+\frac{1}{2}$ or $-\frac{1}{2}$.

This skill is illustrated by Example 8.5.

7. Given the position of an element in the Periodic Table, write its outer electron configuration.

Figure 8.9 is useful in reminding you of which sublevels are being filled. More specifically, the group number tells you the total number of electrons in this outermost principal energy level. The number of the period (horizontal row) tells you the value of n itself. Example 8.6 illustrates this skill.

- -

Arsenic is located in the fourth period and in Group 5. What is the electron configuration for the highest principal energy level? _____

Given the period, n = 4. From the group number, you know that there are a total of 5 electrons in this fourth principal level. Knowing that 2 electrons fill an s sublevel; 6, a p sublevel; 10, a d sublevel; etc., this fourth level must be $4s^2 4p^3$.

- -

This is probably the most useful skill to master in this chapter. For additional practice, see the catalog of problems. Note that in writing the outer configurations for the transition metal atoms, you should first be able to write those for the atoms of the first row (atomic numbers 21–30). Any others you would be able to do by analogy. Example: Given Ti as having $3d^2 4s^2$ would suggest that the element directly below, Zr, has $4d^2 5s^2$. (It turns out that exceptions are often the rule among the transition metals. Don't be too concerned about it at this point.)

8. Use the Periodic Table to predict relative values of atomic radius and ionization energy.

A typical prediction of this type is shown in Example 8.7. In predicting how properties vary with position in the Periodic Table, you may find the following diagram helpful. Properties are nearly constant along the diagonal line D.

Periodic Table

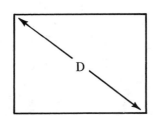

large atomic radius
low ionization energy

small atomic radius
high ionization energy

SELF-TEST

True or False

    1. The atomic number of an element is equal to the number of ( )
electrons in a neutral atom of that element.

    2. When an atom absorbs a photon of energy E, the atom ( )
undergoes an increase in energy equal to or less than E.

    3. The greater the difference in energy between two levels, the ( )
longer the wavelength of the light emitted or absorbed when an
electron moves between them.

    4. The energy carried by photons decreases in the order ( )
infrared > visible > ultraviolet.

    5. The effect of Hund's rule is to give the maximum number of ( )
electrons with parallel spins.

Multiple Choice

    6. The total number of electrons that may occupy the ( )
principal energy level **n** is
      (a) 2   (b) 8
      (c) n   (d) $2n^2$

    7. For an electron with quantum number $\ell = 2$, the quantum ( )
number m_ℓ can have
      (a) only one value   (b) any one of 3 values
      (c) any one of 5 values   (d) an infinite number of
                                         values

8. The number of electrons which can be accommodated in an ( )
$\ell = 2$ sublevel is
(a) 2   (b) 6
(c) 10   (d) 14

9. The possible values of the quantum number m_ℓ for a 3p ( )
electron are
(a) 0, 1, 2   (b) 1, 2, 3
(c) 1, 0, -1   (d) 2, 1, 0, -1, -2

10. The element whose neutral, isolated atoms have three ( )
half-filled 2p orbitals is
(a) $_5$B   (b) $_6$C
(c) $_7$N   (d) $_8$O

11. The electronic structure of Ca (atomic number 20) is ( )
(a) $1s^2 1p^6 1d^{10} 1f^2$ (b) $1s^2 2s^2 2p^6 3s^2 3p^6 3d^2$
(c) $1s^2 2s^2 2p^6 3s^2 3p^6 4s^2$ (d) $1s^2 2s^2 2p^6 3s^2 3p^6$

12. The electron configuration of Fe is $[_{18}Ar]$ $4s^2 3d^6$. The ( )
number of unpaired electrons on the orbital diagram of Fe is
(a) 0   (b) 2
(c) 4   (d) 6

13. Which of the following represents a reasonable set of ( )
quantum numbers for a 3d electron?
(a) 3, 2, 1, ½   (b) 3, 2, 0, -½
(c) neither of these   (d) both of these

14. The most stable arrangement for the outer electrons in the ( )
boron, $_5$B, atom looks like
(a) (↑↓) (↑)( )( )   (b) (↑) (↑)(↑)( )
(c) ( ) (↑)(↑)(↑)   (d) ( ) (↑↓)(↑)( )

15. Which of the following could not be the orbital diagram for ( )
an atom in its ground state?

	1s	2s	2p	3s
(a)	(↑↓)	(↑↓)	(↑↓) (↑↓) (↑↓)	(↑)
(b)	(↑↓)	(↑↓)	(↑↓) (↑↓) (↑↓)	(↑↓)
(c)	(↑↓)	(↑↓)	(↑↓) (↑) (↑)	
(d)	(↑↓)	(↑↓)	(↑↓) (↑↓) ( )	

16. The group number for an element in the Periodic Table ( )
gives the
(a) effective nuclear charge
(b) number of outer electrons
(c) number of elements in the group
(d) value of n for the outer electrons

17. In Group 8 of the Periodic Table, the first ionization energy ( )
decreases with increasing atomic number. This is explained, at least
in part, by
    (a) increasing atomic mass
    (b) increasing nuclear charge
    (c) increasing atomic radius
    (d) increasing boiling point

18. In which of the following series are the atoms arranged in ( )
order of increasing ionization energy?
    (a) Li, Na, K   (b) B, Be, Li
    (c) O, F, Ne   (d) C, P, Se

19. Atomic radius decreases ( )
    (a) within a group, from top to bottom
    (b) within a period, from left to right
    (c) as electrons enter higher principal energy levels
    (d) with more effective shielding of nuclear charge

20. Experimental support for the details of electronic structure ( )
involve measurements of
    (a) wavelengths of atomic spectra
    (b) magnetic properties
    (c) ionization energies
    (d) all of these and more

Problems

1. For an atom of phosphorus, $_{15}P$:
    (a) write the complete electronic configuration.
    (b) give the orbital diagram for the outer principal energy level.

2. Consider the oxygen atom, $_8O$. For each electron in the sublevel of
highest energy, write a complete set of four quantum numbers.

3. Consulting only a Periodic Table, give outer-level configurations
for:
    (a) the atom with the larger radius, $_{11}Na$ or $_{12}Mg$
    (b) $_{23}V$
    (c) $_{35}Br^-$, assuming all electrons have as low energy as possible

4. The electronic energy of the hydrogen atom is given by the
expression:

$$E = \frac{-1312}{n^2} \frac{kJ}{mol}$$

For the transition **n** = 3 to **n** = 1, determine:

(a) ΔE, in kilojoules per mole

(b) λ, in nanometres: $\Delta E = \dfrac{1.196 \times 10^5}{\lambda} \dfrac{kJ \cdot nm}{mol}$

(c) Is light absorbed or emitted? What region of the spectrum is it in?

    5. Calculate the temperature of H(g) at which the average translational energy, $3RT/2$, would be just equal to the energy required to raise the electron to its first excited state (**n** = 2). (R = 8.31 J/mol·K)

SELF-TEST ANSWERS

1. **T** (Recall the discussion in Chapter 2. The atomic number also gives the number of protons and the nuclear charge.)
2. **F** (Equal to, but neither more nor less than.)
3. **F** (Wavelength is inversely proportional to energy. See Equation 8.1.)
4. **F** (Reverse the order.)
5. **T** (Quoted from the text, Section 8.5.)
6. **d**
7. **c** (Corresponding to five geometric orientations of these d
    c orbitals.)
8. **c** (Each of five d orbitals may hold 2 electrons.)
9. **c** (Corresponding to the three orientations of the p orbitals.)
10. **c** (With the outer-level diagram (↑↓)   (↑) (↑) (↑).)
11. **c**
12. **c** (Write it out. Note that $[_{18}Ar]$ is an abbreviation for the electron configuration of argon.)
13. **d**
14. **a** (2s fills before any electrons are added to 2p.)
15. **d** (Recall Hund's rule.)
16. **b** (Choice a is not very far from the truth.)
17. **c** (Along with more shielding of the nuclear charge.)
18. **c** (Note that C, P, and Se, on the diagonal, have very similar ionization energies.)
19. **b** (Larger nuclear charge, with little additional shielding, contracts the electron cloud.)
20. **d** (More would include numerous properties related by the Periodic Table.)

Solutions to Problems

1. (a) $1s^2 2s^2 2p^6 3s^2 3p^3$ or $[_{10}Ne] 3s^2 3p^3$
    (b)   3s   3p
          (↑↓) (↑)(↑)(↑)

2. The complete configuration is $1s^2 2s^2 2p^4$. The electrons considered are $2p^4$. One set of possible quantum numbers:

	n	ℓ	m_ℓ	m_s
1st electron	2	1	1	$+\frac{1}{2}$
2nd electron	2	1	0	$+\frac{1}{2}$
3rd electron	2	1	- 1	$+\frac{1}{2}$
4th electron	2	1	1	$-\frac{1}{2}$

Note that there is no other set of values possible for n and ℓ for these electrons. For three of the electrons, m_s must have the same value. Again, there must be three different values for m_ℓ, and two which are the same.

3. (a)   The Mg has the larger nuclear charge with little additional shielding, which makes it the smaller atom.
          Na: $3s^1$ (Period 3 means n = 3; Group 1 means 1 outer electron.)
    (b) $_{23}V$: $3s^2 3p^6 4s^2 3d^3$ (Period 4 means n = 4. But since there is some "overlapping" between 3d and 4s, we have given all this.)
    (c) Br^-: $4s^2 4p^6$ (Period 4 and Group 7 imply $4s^2 4p^5$. Note that the indicated charge means that there is an extra electron.)

4. (a)   For n = 1: $E_1 = -1312/(1)^2 = -1312$ kJ/mol
          n = 3: $E_3 = -1312/(3)^2 = -146$ kJ/mol
          So, $\Delta E = E_1 - E_3 = -1312 + 146 = -1166$ kJ/mol
    (b) $\lambda = \dfrac{1.196 \times 10^5}{/\Delta E/}$ nm = $\dfrac{1.196 \times 10^5}{1166}$ = 102.6 nm
    (c)   Ultraviolet light is emitted, carrying off the energy lost by the atoms.

5. The energy required is calculated as in Problem 4:
   $\Delta E = E_2 - E_1 = -1312/(2)^2 + 1312/(1)^2 = 984$ kJ/mol
   $$T = \frac{2E}{3R} = \frac{2(984 \times 10^3 \text{ J/mol})}{3(8.31 \text{ J/mol·K})} = 7.89 \times 10^4 \text{ K}$$

SELECTED READINGS

Alternative discussions of electron structure are given in the following paperbacks:

Hochstrasser, R.M., *Behavior of Electrons in Atoms: Structure, Spectra, and Photochemistry of Atoms,* New York, W.A. Benjamin, 1964.

Pimentel, G.C., *Chemical Bonding Clarified through Quantum Mechanics,* San Francisco, Holden-Day, 1969.

Sisler, H.H., *Electronic Structure, Properties, and the Periodic Law,* New York, Van Nostrand, 1973.

Mainly of historical or philosophical interest:

Bohr, N., On the Constitution of Atoms and Molecules, *Philosophical Magazine, 26* (sixth series; July 1913), pp. 1–25.

d'Espagnat, B., The Quantum Theory and Reality, *Scientific American* (November 1979), pp. 158–181.

Lagowski, J.J., *The Structure of Atoms,* Boston, Houghton Mifflin, 1964.

Lewis, G.N., *Valence and the Structure of Atoms and Molecules,* New York, Dover, 1966 (reprint).

For a look at the experimental support for configuration theory, and an extension to molecular structure:

Hänsch, T.W., The Spectrum of Atomic Hydrogen, *Scientific American* (March 1979), pp. 94–110.

Sanderson, R.T., Ionization Energy and Atomic Structure, *Chemistry* (May 1973), pp. 12–15.

Wahl, A.C., Chemistry by Computer, *Scientific American* (April 1970), pp. 54–66.

Zare, R.N., Laser Separation of Isotopes, *Scientific American* (February 1977), pp. 86–98.

9

GROUPS 1 AND 2; THE METALS AND THEIR COMPOUNDS

QUESTIONS TO GUIDE YOUR STUDY

1. What are the relative abundances, at the surface of the earth, of the metals of Groups 1 and 2?

2. How do the metals occur in nature? What are the principal ores of commercial value? Are they the metals themselves, or oxides, sulfides, or other compounds?

3. By what reactions are the metals extracted from their ores?

4. What properties are characteristic of metals? Are the metals of Groups 1 and 2 typical metallic substances?

5. What reactions do the metals themselves take part in? What are some of the properties and reactions of their simple compounds with nonmetals? What are the major uses for these substances?

6. What regularities and trends have you seen among these elements in their atomic properties (e.g., atomic radius and ionization energy)? What explanations have been given in terms of electronic structure? What correlations made with position in the Periodic Table?

7. Are there similarities and trends to be seen in the thermochemistry of the reactions of the metals? Are there similarities in the crystal structures of the metals and their compounds?

8. What is the nature of the attractive forces in a bulk sample of the metals; in a sample of a crystalline compound like $NaCl$?

9. How is a simple and useful compound, say Na_2CO_3, prepared commercially? How would you prepare it in the laboratory?

10. What are some of the differences between the metals, and their compounds, of Group 1 and Group 2?

YOU WILL NEED TO KNOW

This chapter applies the principles of the preceding chapters, with special emphasis on the concepts and calculations introduced in Chapters 6 to 8. It would also be useful to review the discussion of the sources, preparations, and uses for sodium and magnesium (Section 5.3 in the text).

Concepts

1. What is meant by atomic radius, ionization energy, and atomic spectra. See Chapter 8.

2. How to write the electronic configuration, particularly for the outermost principal energy level, for any major group element. See Basic Skill 7 in Chapter 8 of this guide.

3. How to recognize the formulas and names of common ions. See Table 3.2 and Figure 3.3 in the text.

Math

1. How to relate the energy change for an atom and the wavelength for a line in its spectrum. See Basic Skill 1 in the preceding chapter.

2. How to use Hess's Law. In addition, how to relate amounts of reacting species to the heat flow in a reaction. How to perform calorimetric calculations. See Basic Skills 3 to 5 in this guide, Chapter 6.

3. How to write balanced equations for chemical reactions, including the formation of acidic and basic solutions — Chapter 4.

4. How to use the Ideal Gas Law to calculate one variable, given all others. How to relate the volume of a gaseous reactant or product to the mass or moles of substances involved in a reaction. See Basic Skill 2(b) in Chapter 7 of this guide.

5. How to relate the dimensions (edge, body and face diagonals) of a cube. For a cube of edge s, face diagonal $= \sqrt{2}\,s$; body diagonal $= \sqrt{3}\,s$. The simple math involved here, use of the Pythagorean theorem, is reviewed in this chapter of the text.

BASIC SKILLS

Many of the skills used in this chapter were discussed earlier in this guide. You should become thoroughly familiar with these skills, if you are not already. A review is always helpful in mastering a subject. The skills reviewed include the following:

— *Relate the wavelength of a spectral line to the change in energy of an atom.* Example 9.1 (as well as 8.1) illustrates this.

— *Calculate* ΔH *for a reaction from heats of formation.* Example 9.4 (as well as 6.3 and 6.4) uses data from Tables 6.1 and 6.2 to do this. The accompanying exercise carries the calculation a step further, calling for the use of the calorimetric relations:

$$Q_{reaction} = -Q_{water} \quad \text{and} \quad Q_{water} = 4.18 \ \frac{J}{g \cdot °C} \times m_{water} \times \Delta t.$$

(Also see Example 6.6 .)

— *Solve the Ideal Gas Law for one variable, given the values of the others.* Both Examples 9.5 and 9.6 require the calculation of V after first having calculated n based on a balanced equation. (An earlier example of this: Example 7.7.)

— *Write and balance simple chemical equations.* Example 9.7 considers a problem entirely analogous to one described in the body of the text.

Skills new to this chapter include the following.

1. Write balanced chemical equations for the reactions of a metal of Group 1 or 2 with a halogen (Group 7 element), N_2, O_2, or H_2O.

Table 9.2 summarizes all the possible reactions and includes two additional types: reaction with H_2 and with sulfur. Example 9.2 calls for equations involving Ca and a halogen, N_2, and O_2. Example 9.5 requires that an equation first be written for the reaction of K and O_2 to give the peroxide, K_2O_2.

Note that in reactions involving O_2, three products are possible: the metal oxide, peroxide, or superoxide. Again, Table 9.2 indicates which product is to be expected, depending on the metal reactant. You should realize that all the metals and all the compounds of the metals with the nonmetallic elements are solids at 25°C and 1 atm. Knowing the charges on the ions in these compounds (+1 for a Group 1 metal, +2 for a Group 2 metal, −1 for a halide, etc.) should allow you to easily write the formulas. (See Skill 5 in Chapter 3 of this guide.)

See the text problems for additional practice in writing formulas and equations involving these metals. Also see Equations 9.2 to 9.11 in the body of the text.

2. Given sufficient thermodynamic data, determine the lattice energy of an ionic compound.

Example 9.3 illustrates the straightforward application of Hess's Law to such a calculation. The catalog of problems includes additional practice in this skill. Such thermodynamic analysis proves to be helpful in understanding many other kinds of reactions too.

3. Given the type of cubic unit cell (simple, body-centered, face-centered), relate the cell dimensions to atomic radius.

In its simplest form, this skill is used to work Example 9.8 and the accompanying exercise. Example 9.9 extends the skill to the determination of ionic radii. In each case, you must know which atoms or ions are in contact and along which cell dimension (edge, face or body diagonal) contact is made. Additional crystal structure calculations are required in the problems at the end of the chapter.

4. Describe some of the sources, preparations, and uses of the Group 1 and 2 metals and their compounds.

A lot of this information is summarized in the tables of this chapter: 9.1 (physical properties of the metals); 9.2, referred to above (reactions of the metals); 9.3 (solubilities of ionic compounds of these metals). Additional discussion is presented concerning the chemistry and uses of such compounds as $NaCl$, $NaOH$, Na_2CO_3, and $NaHCO_3$, and selected calcium compounds (the oxide, hydroxide, carbonate, and sulfate). Reaction types you have seen before and will see again in this discussion include the formation of acidic and basic solutions, reactions of acids and bases, precipitation, and oxidation-reduction.

As with the descriptive material of Chapter 5, this chemistry will be learned through repeated practice in writing equations, working problems, explaining properties and their trends in terms of what has come before, and finally through some memorization. Just how much you should be able to recall is up to you and your instructor. For that most useful practice, do more problems in the text.

SELF-TEST

True or False

1. The metals of Group 1 are generally less reactive than those of Group 2.   (   )

2. Simple ionic compounds of Group 2 metals are generally less soluble in water than those of Group 1.   (   )

3. For any atom, the second ionization energy is larger than the first.   (   )

4. The +2 ion is much larger than the neutral magnesium atom.   (   )

5. A compound of magnesium (Group 2) and sulfur (Group 6) ( )
is expected to have the formula MgS.

6. A zeolite column used to soften water could be regenerated ( )
by washing it with a concentrated solution of $CaCl_2$.

7. The lattice energy of a Group 1 metal halide is large and ( )
negative.

Multiple Choice

8. Within a group in the Periodic Table, metallic character ( )
   (a) increases with increasing atomic number
   (b) decreases with increasing atomic number
   (c) remains more or less constant throughout
   (d) increases with increasing ionization energy

9. The electron configuration of any Group 2 metal ion is ( )
   (a) ns^1 (b) ns^2
   (c) $ns^2 np^2$ (d) ns^2 or $ns^2 np^6$

10. In which of the following series are the ions arranged in ( )
order of increasing radius?
   (a) $K^+ < Na^+ < Mg^{2+}$ (b) $Na^+ < K^+ < Mg^{2+}$
   (c) $Mg^{2+} < Na^+ < K^+$ (d) $Mg^{2+} < K^+ < Na^+$

11. The molar heats of formation of NaF, NaCl, and NaBr are ( )
$-569, -411$, and -360 kJ. That of NaI is likely to be
   (a) more negative than -569 kJ
   (b) more positive than -360 kJ
   (c) -447 kJ, the average
   (d) unpredictable

12. $K_2 SO_4$ is a common ionic substance. The anion has the ( )
formula
   (a) K^+ (b) K_2^{2+}
   (c) S^{2-} (d) SO_4^{2-}

13. A water solution of $Na_2 CO_3$ might be expected to contain ( )
   (a) Na^+ (b) CO_3^{2-}
   (c) $H_2 O$ (d) all of these species

14. Which of the following species may "harden" water? ( )
   (a) Na^+ (b) Mg^{2+}
   (c) OH^- (d) CO_2

15. The number of atoms in contact with the one at the cube   (   )
center in a body-centered cubic cell is
    (a) 2   (b) 6
    (c) 8   (d) 12

16. A certain metal crystallizes with a face-centered cubic cell.   (   )
The relationship between the atomic radius (r) and the edge (s) of
the cell is:
    (a) $2r = s$ (b) $2r = \sqrt{2}\,s$
    (c) $4r = \sqrt{2}\,s$ (d) $4r = \sqrt{3}\,s$

17. What is the principal commercial source of sodium?   (   )
    (a) Na_2O (b) NaOH
    (c) Na_2CO_3 (d) NaCl

18. In the preparation of sodium carbonate by the Solvay   (   )
process, the reactants in the overall process are
    (a) Na and CO_2 (b) Na, H_2, and CO_2
    (c) $CaCO_3$ and NaCl   (d) NH_3 and $CaCl_2$

19. To prepare hydrogen peroxide in the laboratory, you might   (   )
add _____ to water.
    (a) $\frac{1}{2}O_2$ (b) Na
    (c) Li_2O (d) Na_2O_2

20. The commercial preparation of NaOH involves   (   )
    (a) separation of NaOH from seawater
    (b) direct reaction of the elements
    (c) electrolysis of NaCl(aq)
    (d) electrolysis of NaCl(1)

Problems

1. Write balanced equations for the reactions that take place when
    (a) potassium metal is added to water
    (b) sodium comes into contact with air
    (c) a solution of NaOH is left open to the air
    (d) limestone is heated to $1000°C$
    (e) a water solution of $Ca(OH)_2$ is deionized

2. Given the following data:

	ΔH_f (kJ/mol)
$Na^+(aq)$	-239.7
$OH^-(aq)$	-229.9
$H_2O(1)$	-285.8

Calculate ΔH when 1.00 g of sodium (atomic mass = 23.0) reacts with water:

$$2\ Na(s) + 2\ H_2O(1) \rightarrow 2\ Na^+(aq) + 2\ OH^-(aq) + H_2(g)$$

3. Calculate the volume of hydrogen gas formed at 24°C and 720 mm Hg (96.0 kPa) for the reaction in Problem 2. (R = 0.0821 $\ell\cdot$atm/(mol$\cdot$K) or, in SI, R = 8.31 kPa$\cdot$dm^3/(mol$\cdot$K)).

4. Strontium (atomic mass = 87.62) crystallizes in a face-centered cubic structure with the edge of the unit cell being 0.608 nm.
   (a) Make a sketch of a face of the unit cell.
   (b) Determine the atomic radius of strontium.

5. Lithium iodide crystallizes with the structure of NaCl. The shortest Li–I internuclear distance has been determined experimentally (by x-ray diffraction) to be 0.302 nm.
   (a) Calculate the shortest I–I distance.
   (b) Assuming that the distance calculated in (a) is the diameter of the iodide ion in LiI, what can be said about the size of the lithium ion (Li^+)? The tabulated radius for Li^+ is 0.060 to 0.068 nm, depending on where you look. Is it consistent with your conclusions?

SELF-TEST ANSWERS

1. **F** (More so. Why? A major reason is the greater ease of a Group 1 metal losing its outermost electron.)
2. **T** (See Table 9.3. Why do you suppose this is so? More on solubility in Chapter 16.)
3. **T** (Given the positive charge, a second electron is more difficult to remove.)
4. **F** (A cation is smaller than the parent neutral atom. Electrons have been removed and those that remain are pulled in more closely by the net positive charge.)
5. **T** (The expected ions with noble-gas configurations are Mg^{2+} and S^{2-}.)
6. **F** (Use NaCl so as to be able to exchange Na^+ for any Ca^{2+} in hard water.)
7. **T** (As noted in the discussion in Section 9.2, the large, negative lattice energy makes the Group 1 halides stable.)
8. **a** (Metallic character increases as you move to the left and downward.)
9. **d** (ns^2 would apply to Be^{2+}; $ns^2 np^6$ would apply to all the other Group 2 *ions*.)

10. c   (From Mg^{2+} to Na^+, effective nuclear charge decreases; from Na^+ to K^+, outer electrons are in a higher energy level farther from the nucleus.)

11. b   (As expected from the observed trend and correlation with position in the Periodic Table.)

12. d   (Recall Table 3.2.)

13. d

14. b   (Mg^{2+} and Ca^{2+} harden water, forming precipitates with soap and leaving deposits on evaporation.)

15. c   (There are eight corners. Contact is made along the body diagonal.)

16. c   (Contact is made along a face diagonal. See Example 9.8.)

17. d   (Sources of the element were discussed in Chapter 5.)

18. c   (See the net reaction as described by Equation 9.18.)

19. d   (With Na, you would form H_2 and NaOH; with Li_2O, the product is LiOH.)

20. c   (As discussed in Chapter 5 and reviewed here.)

Solutions to Problems

1. (a) $2 K(s) + 2 H_2O(1) \rightarrow 2 K^+(aq) + 2 OH^-(aq) + H_2 (g)$

         (Table 9.2)

    (b) $2 Na(s) + O_2 (g) \rightarrow Na_2 O_2 (s)$ (Table 9.2)

    (c) $2 OH^-(aq) + CO_2 (g) \rightarrow H_2 O + CO_3^{2-}(aq)$ (Equation 9.14)

    (d) $CaCO_3 (s) \rightarrow CaO(s) + CO_2 (g)$

    (e) $Ca^{2+}(aq) + 2 HR(s) \rightarrow CaR_2 (s) + 2 H^+(aq)$

        $2 H^+(aq) + 2 OH^-(aq) \rightarrow 2 H_2 O$ (Example 9.7)

2. For the equation as written:

    $\Delta H = 2(-239.7 \text{ kJ}) + 2(-229.9 \text{ kJ}) + 2(285.8 \text{ kJ}) = -367.6 \text{ kJ}$

    For 1.00 g of Na:

$$\Delta H = 1.00 \text{ g Na} \times \frac{-367.6 \text{ kJ}}{46.0 \text{ g Na}} = -7.99 \text{ kJ}$$

3. $n_{H_2} = 1.00 \text{ g Na} \times \dfrac{1 \text{ mol } H_2}{46.0 \text{ g Na}} = 0.0217 \text{ mol } H_2$

$$V = \frac{nRT}{P} = \frac{0.0217 \text{ mol} \times 0.0821 \frac{\ell \cdot atm}{mol \cdot K} \times 297 \text{ K}}{720/760 \text{ atm}} = 0.558 \, \ell$$

$$V = \frac{nRT}{P} = \frac{0.0217 \text{ mol} \times 8.31 \frac{kPa \cdot dm^3}{mol \cdot K} \times 297 \text{ K}}{96.0 \text{ kPa}} = 0.558 \text{ dm}^3$$

4. (a) Showing atoms in contact along the face diagonal:

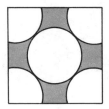

(b) $4r = \sqrt{2}\,s = \sqrt{2}\,(0.608)$
    $r = 0.215$ nm
5. (a) Distance (I–I) = $0.302 \times \sqrt{2} = 0.427$ nm
   (b) If $r\ I^- = 0.214$ nm, then $r\ Li^+$ cannot exceed the value calculated from $r\ I^- + r\ Li^+ \leqslant 0.302$ nm
   That is, $r\ Li^+$ is less than or equal to 0.088 nm. Apparently, Li^+ ions do not touch I^- ions.

SELECTED READINGS

An application is described in:

Probstein, R.F., Desalination, *American Scientist* (May–June 1973), pp. 280–293.

The nature of the solid state and additional descriptive chemistry:

Fullman, R.J., The Growth of Crystals, *Scientific American* (March 1955), pp. 74–80.
Holden, A., *The Nature of Solids,* New York, Columbia University Press, 1965.
Moore, W.J., *Seven Solid States: An Introduction to the Chemistry and Physics of Solids,* New York, W.A. Benjamin, 1967.
Rochow, E.G., *Modern Descriptive Chemistry,* Philadelphia, W.B. Saunders, 1977.
Sanderson, R.T., The Nature of "Ionic Solids," *Journal of Chemical Education* (September 1967), pp. 516–523.

Some unexpected(?) properties of metals:

Chaudhari, P., Metallic Glasses, *Scientific American* (April 1980), pp. 98–117.
Dye, J.L., Anions of the Alkali Metals, *Scientific American* (July 1977), pp. 92–105.

Visualizing the third dimension and experimentally probing it:

Bragg, L., X-Ray Crystallography, *Scientific American* (July 1968), pp. 58–70.
Kapechi, J.A., An Introduction to X-Ray Structure Determination, *Journal of Chemical Education* (April 1972), pp. 231–236.
Wells, A.F., *The Third Dimension in Chemistry,* New York, Oxford University Press, 1956.

10

COVALENT BONDING; MOLECULAR SUBSTANCES

QUESTIONS TO GUIDE YOUR STUDY

1. Why do bonds form between neutral atoms as well as between ions? (Why don't all the atoms in the universe bond together in one super molecule?)

2. Is there anything common to all "types" of bonds, whether ionic, covalent, metallic, etc?

3. How do you account for the observed differences in the strengths of bonds?

4. How are the size and shape of a molecule related to the electronic structures of the component atoms? Is there a simple, reliable way of predicting the geometry?

5. Can you predict formulas of molecular substances from a theoretical basis? In particular, in terms of the electronic structures of the atoms involved?

6. How would you experimentally determine bond energies, lengths, angles, and polarities?

7. How would you account for the fact that an atom may bond to just one other atom, sometimes to two others, or three . . . (e.g., C in CO, CO_2, CH_4)? Is the number of bonds predictable?

8. Can you now give a more detailed description as to what occurs at the atomic-molecular level when a chemical reaction takes place?

YOU WILL NEED TO KNOW

Concepts

1. How to write electron configurations for the outermost energy level of the main-group elements (say, for elements of atomic number 1 through 10). Others you should be able to do by analogy, based on position in the Periodic Table. See Skill 7 in Chapter 8 of this guide.

2. The shapes and relative sizes of atomic orbitals. See the figures of Chapter 8 in the text.

3. What is meant by bond energy. See Chapter 6.

Math

1. A familiarity with the geometry (symmetry and angles) of several plane and solid figures, particularly the tetrahedron, would be helpful. See the figures in Chapter 10 of the text. A molecular model kit would be very useful in visualizing the geometric consequences of bonding. See your instructor, or bookstore, about borrowing or purchasing such a kit.

BASIC SKILLS

1. **Given a Periodic Table, or a table of electronegativities, compare the polarities of different bonds.**

Example 10.1 illustrates this skill. You should know how electro-negativity changes with position of an element in the Periodic Table. It increases as you move left to right and from bottom to top. The most electronegative elements lie at the top right; the most electropositive, bottom left.

Given a table of electronegativity values, you should also be able to estimate the partial ionic character by using Figure 10.4.

Bond polarity is further illustrated by problems at the end of the chapter. See the catalog listing.

2. **Draw Lewis structures for molecules and polyatomic ions.**

This skill is required for Examples 10.2 and 10.3. A majority of the text problems also require its use, as do the next four skills below.

We cannot emphasize too strongly that the ability to draw Lewis structures is fundamental to a great many other skills in this chapter. You cannot predict the geometry of a molecule, decide whether or not it is polar, or indicate the type of hybrid orbitals used, among other things, unless you know the Lewis structure.

Sometimes, more than one plausible Lewis structure can be written for a given molecule or ion. Such structures usually differ only in their skeletons — that is, in the order in which atoms are bonded to one another. This point is discussed in Example 10.2, which provides a few rules to guide you to the correct skeletal structure. For example, carbon ordinarily forms four bonds, while hydrogen forms only one. Again, oxygen atoms in a polyatomic species normally bond to a central nonmetal atom. As you progress in your study of chemistry, you will gradually add to your store of rules of this type.

3. **Given or having written the Lewis structure of a molecule or ion, predict its geometry.**

The reasoning here is shown in Example 10.4 (only single bonds) and Example 10.5 (species containing double bonds). There are two principles involved:

(a) The electron pairs around a given atom lie in orbitals which are directed as far apart from each other as possible. Thus, two electron pairs are oriented at $180°$ to each other, along a straight line, and in opposite directions. Three pairs are directed toward the corners of an equilateral triangle. Four pairs are directed toward the corners of a regular tetrahedron.

(b) So far as geometry is concerned, a multiple bond behaves as if it were a single pair of electrons. So, in SO_2, for example, we ignore the extra pair in the double bond and count only three pairs of electrons around the central sulfur atom to arrive at a predicted bond angle of $120°$.

Note that you must know the Lewis structure before you can predict the geometry. Note, too, that in describing molecular geometry, we indicate only the positions of the atoms. The location of any unshared electron pairs is not included in the description. So, for example, NH_3 is a pyramidal molecule and not a tetrahedral one.

Table 10.5 summarizes all the possible electron arrangements and geometries considered. See the catalog of problems for additional practice in this important skill.

4. **Given or having predicted the geometry of a molecule or ion, predict whether or not it is polar.**

The general principle here is a simple one: a molecule or polyatomic ion can be nonpolar only if

(a) all the bonds are nonpolar, as in H_2 or P_4.

(b) the various polar bonds cancel each other, as would be the case in the symmetrical species

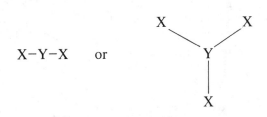

If the symmetry is destroyed by substituting an atom other than X as in

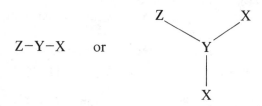

$$Z-Y-X \quad \text{or}$$

or by changing the bond angle as in

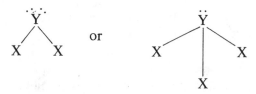

then the species is polar. This principle is illustrated in Example 10.6 and in a few of the problems at the end of the chapter. Again, note that you must first know the geometry before you can predict the polarity. Geometry can be predicted only if the Lewis structure is known.

5. **Given the Lewis structure of a molecule or ion, indicate the hybrid orbitals used by the central atom.**

The hybrid orbitals discussed in this chapter include sp (two orbitals directed away from each other at 180°), sp² (three, directed at 120°), and sp³ (four, directed at 109.5°). These are the orbitals that determine the overall geometry of a species and so include none of the multiple bond electron pairs (which you were told above to ignore.)
Example 10.7 uses this skill. It is also required in some of the problems at the end of the chapter.

6. **Indicate the kind and number of bonds (sigma and pi) in a species, given its Lewis structure.**

Example 10.8 illustrates this skill. As can be seen from this example, any single bond we consider is a sigma bond. Any double or triple bond consists of a sigma bond (electrons in a hybrid orbital) plus one or two pi bonds (electrons in nonhybridized orbitals), respectively.

SELF-TEST

True or False

1. Chemical bonds almost never form unless half-filled orbitals   ( )
are available.

2. Bond energy increases in the order single bond < double   ( )
bond < triple bond.

3. So far as geometry is concerned, a multiple bond behaves as   ( )
if it were a single pair of electrons.

4. A binary compound of an element having a low ionization   ( )
energy and an element having a high electronegativity is likely to
possess covalent bonds.

5. The Lewis structure for any Group 3 atom would be drawn   ( )
·X·.

6. There are 24 valence electrons in the $SO_3{}^{2-}$ ion.   ( )

7. Experimental results support the idea that a certain mole-   ( )
cule AB_2 is linear. This must mean that there are no unshared
electron pairs on the central atom.

Multiple Choice

8. Which one of the following contains both ionic and covalent   ( )
bonds?
    (a) NaOH   (b) HOH
    (c) CH_3Cl (d) P_4

9. Which one of the following species would have an unpaired   ( )
electron?
    (a) SO_2 (b) $NO_2{}^-$
    (c) $NO_2{}^+$ (d) NO_2

10. Which one of the following bonds would be the least polar?   ( )
    (a) H−F   (b) O−F
    (c) Cl−F   (d) Ca−F

11. The molecular shape of PH_3 is predicted as being   ( )
    (a) linear   (b) planar
    (c) pyramidal   (d) tetrahedral

12. sp^3 hybridization is important in describing the bonding in   (   )
   (a) H_2O (b) CH_4
   (c) NH_4^+ (d) all of these

13. What type of hybridization is shown by carbon in the CH_2O (   )
molecule, which has the Lewis structure:

$$H \diagdown$$
$$C=\ddot{\underset{..}{O}}$$
$$H \diagup$$

   (a) sp   (b) sp^2 (c) sp^3 (d) sp^4

14. The weakest N to N bond among the following is that in   (   )
   (a) $:N\equiv N:$ (b) $F-\ddot{N}=\ddot{N}-F$
   (c) $F-\ddot{N}-\ddot{N}-F$ (d) $H-\ddot{N}=N=\ddot{N}:$
            | |
            F F

15. A bond angle of $120°$ appears in species with which geom-   (   )
etry?
   (a) linear   (b) square
   (c) triangular   (d) tetrahedral

16. Consider the following molecules: (   )

$$\begin{matrix} & & H\ H & F\ F \\ & & |\ \ | & |\ \ | \\ F-C\equiv C-F & H-C\equiv C-F & F-C=C-F & F-C-C-F \\ & & & |\ \ | \\ & & & H\ F \end{matrix}$$

In how many of these molecules are there two pi bonds between the
carbon atoms?
   (a) 1   (b) 2   (c) 3   (d) 4

17. An observed bond angle of about $107°$ would probably be   (   )
explained in terms of _____ hybridization.
   (a) no   (b) sp
   (c) sp^2 (d) sp^3

18. Which one of the following species does not obey the octet   (   )
rule?
   (a) BeF_2 (b) NO_2
   (c) H_2 (d) all of these

19. What is the number of valence electrons in the ClO_2^- ion?   (   )
   (a) 18   (b) 19
   (c) 20   (d) 21

20. The skeleton structure of the ClO_2^- ion referred to in (   )
Question 19 is O–Cl–O. How many valence electrons are left to add
to the skeleton structure to give the Lewis structure of ClO_2^-?
   (a) 14   (b) 16
   (c) 18   (d) 20

Problems

1. Draw a Lewis structure for each of the following species:
   (a) CS_2
   (b) NO_2^-
   (c) H_3O^+

2. Predict the bond angle (109.5°, 120°, or 180°) for each species,
given the Lewis structure.
   (a) NO_2^+,   :Ö=N=Ö:

   (b) SO_3,   :Ö=S–Ö:
                              |
                             :O:

   (c) ClO_3^-,   :Ö–Cl–Ö:
                              |
                             :O:

3. For each species of Problem 2, note whether it is expected to be a
dipole.

4. For each species of Problem 2, indicate the hybridization of the
central atom.

5. The sequence in which the atoms are bonded in the hydrogen
peroxide molecule is H–O–O–H. What does the fact that it is polar tell you
about its geometry?

SELF-TEST ANSWERS

1. F (Note that hybridization often results in the unpairing of
      electrons: consider Be in BeF_2, B in BF_3, and C in CH_4.)
2. T (For example, consider carbon-carbon energies in Table 10.2.)

3. **T** (See the discussion of the electron-pair repulsion principle, Section 10.3.)

4. **F** (Ionic or at least somewhat polar. The combination described is one of a metal and a nonmetal.)

5. **T** (Two of the dots, i.e., electrons, might be shown as paired. See Table 10.4.)

6. **F** (The charge implies the presence of two additional electrons.)

7. **T** (One unshared pair would give a bond angle of about $120°$; two unshared pairs, about $109.5°$. See Table 10.5)

8. **a** (You should recognize the ions Na^+ and OH^-. The anion contains a polar covalent bond.)

9. **d** (There is an odd number of valence electrons.)

10. **b** (The electronegativity difference would be the smallest for two atoms in adjacent positions in the Periodic Table.)

11. **c** (One unshared pair on the central P atom. The geometry is like that of NH_3, and describes the positions of the atoms.)

12. **d** (Each has four separate pairs of electrons on the central atom.)

13. **c** (The Lewis structure shows effectively three pairs of electrons on the carbon. The fourth, in the double bond, does little in determining the geometry.)

14. **c** (A single bond is the weakest. See Question 2.)

15. **c** (Whether two or three of the corners of the triangle are occupied by atoms. If two corners are occupied, the description would be "bent".)

16. **b**

17. **d** (This is close to the tetrahedral angle. For such slight departure in bond angle from that associated with the given hybrid, see the discussion following Example 10.4.)

18. **d** (BeF_2 has only two pairs, four electrons, about the central atom. NO_2 has an odd number of valence electrons. H_2 has only one pair.)

19. **c** (Remember to add one for the minus charge.)

20. **b** ($20 - 4 = 16$.)

Solutions to Problems

1. (a) $:\ddot{S}=C=\ddot{S}:$ (Looks like CO_2. Note that S and O are both in Group 6.)

   (b) $:\ddot{O}=\ddot{N}-\ddot{O}:$

   (c) $H-\ddot{O}-H$
   $\quad\quad\quad |$
   $\quad\quad\quad H$

2. (a) 180° (Effectively only two pairs on the central atom.)
   (b) 120° (Effectively only three pairs on the central atom.)
   (c) 109.5° (Count four pairs on Cl; direct them toward the corners of a tetrahedron.)
3. (a) No (Symmetrical.)
   (b) No (Symmetrical.)
   (c) Yes (Nonsymmetrical.)
4. (a) sp
   (b) sp^2
   (c) sp^3
5. That the two O−H bonds are not directed in space so as to "cancel" each other. (For H_2O_2 to be nonpolar would require that the entire molecule be planar, with O−H bonds directed away from each other.)

SELECTED READINGS

Alternative and extensive discussions of bonding, the theories and their history are given in:

Companion, A.L., *Chemical Bonding,* New York, McGraw-Hill, 1979.
Coulson, C.A., *The Shape and Structure of Molecules,* New York, Oxford, 1973.
Gray, H.B., *Chemical Bonds,* Menlo Park, Calif., W.A. Benjamin, 1973.
Lagowski, J.J., *The Chemical Bond,* Boston, Houghton Mifflin, 1966.
Lewis, G.N., *Valence and the Structure of Atoms and Molecules,* New York, Dover, 1966 (reprint).
Pauling, L., *The Nature of the Chemical Bond,* Ithaca, N.Y., Cornell, 1960.
Pimentel, G.C., *Chemical Bonding Clarified through Quantum Mechanics,* San Francisco, Holden-Day, 1969.
Sisler, H.H., *Electronic Structure, Properties, and the Periodic Law,* New York, Van Nostrand, 1973.

The experimental determination of molecular structure and dynamics is considered in the readings on x-ray diffraction listed in Chapter 9 and:

Barrow, G.M., *The Structure of Molecules: An Introduction to Molecular Spectroscopy,* New York, W.A. Benjamin, 1963.
Brey, W.S., Jr., *Physical Methods for Determining Molecular Geometry,* New York, Reinhold, 1965.
Sonnessa, A.J., *Introduction to Molecular Spectroscopy,* New York, Reinhold, 1966.
Wagner, J.J., Nuclear Magnetic Resonance Spectroscopy — An Outline, *Chemistry* (March 1970), pp. 13–15.

Shapes of molecules, and the use of models:

Hall, S.K., Symmetry, *Chemistry* (March 1973), pp. 14–16.
Morrow, F.J., Do-it-yourself Molecular Models, *Chemistry* (April 1972), pp. 6–9.
Pauling, L., *The Architecture of Molecules,* San Francisco, W.H. Freeman, 1964.

11

LIQUIDS AND SOLIDS

QUESTIONS TO GUIDE YOUR STUDY

1. What properties do you usually associate with liquids and solids? Are there simple laws, as there are for gases, relating volume, temperature, and pressure for a liquid or solid?

2. What kinds of substances normally exist as liquids (or solids or gases) at 25°C and 1 atm? Are there correlations to be made with position in the Periodic Table? Or with molecular-level structure?

3. How would you describe the process of evaporation — in terms of what you would observe; in terms of atomic-molecular behavior?

4. Again, how would you recognize boiling, or sublimation or melting? What would you measure? What energy effects accompany these processes?

5. What kind of experiment would you do to show the presence of a vapor above a liquid? Or to show how the pressure of the vapor changes with temperature?

6. How do you account for the fact that some solids, like dry ice, are converted directly to gases without first melting?

7. What properties are characteristic of ionic solids? What properties characterize molecular and metallic substances? How are they explained in terms of bonding?

8. What factors favor the formation of small, discrete molecules; of large, extensive macromolecules? How would you experimentally distinguish between very small and very large molecules?

9. What kind of experimental evidence would tell you that you were dealing with interatomic forces? With intermolecular forces? What are the relative magnitudes of the energies required to overcome these forces?

10. Can you now explain more fully, in terms of particle structure and the forces between particles, what happens during physical and chemical changes?

YOU WILL NEED TO KNOW

Concepts

1. The meaning of ΔH and the interpretation of its sign; how to use Hess's Law. See Chapter 6.

2. The ideas of kinetic theory about the nature of a gas. (This chapter of the text extends kinetic theory to liquids and solids.) See Chapter 7.

3. How to recognize whether a substance is likely to be ionic, molecular, or metallic. This is a prediction based on position in the Periodic Table and/or differences in electronegativity. See Chapters 1 and 10.

4. How to predict molecular geometry and polarity — Chapter 10.

Math

1. How to use the Ideal Gas Law to calculate one variable from values of all the others — Chapter 7. (See Skill 2(b) in Chapter 7 of this guide.)

BASIC SKILLS

1. Have sufficient understanding of the concept of vapor pressure to work problems such as 11.2–11.4.

This is an awkward way to phrase a skill, but the only alternative would be to resort to a lot of jargon. In certain parts of these problems, you have to apply the Ideal Gas Law to calculate one variable from all the others (usually a pressure or volume is needed). See Example 11.1. You should keep in mind that this law can be applied only when there is a single gaseous phase present. It cannot be used, for example, to calculate the pressure of a vapor in equilibrium with its liquid. That pressure is a characteristic property of a particular liquid at a given temperature and is known as the vapor pressure.

In other problems, you have to use a plot of vapor pressure versus temperature (like Figure 11.3), or a complete phase diagram (see Figure 11.13), to relate the vapor pressure to the temperature. See Example 11.2.

2. Predict and describe the kinds of intermolecular forces present in a given molecular substance.

Example 11.3 illustrates this point in a simple way. Example 11.4 carries the skill farther so as to include all the types of intermolecular forces considered:

(a) dispersion forces, common to all molecular substances;

(b) dipole forces, restricted to polar molecules;

(c) hydrogen bonds, found in polar molecules where H is bonded to F, O, or N.

Note that to use this skill, you must first decide whether the molecule in question is polar or nonpolar. This in turn depends on your knowing the geometry of the molecule. See Skill 4 in Chapter 10 of this guide.

3. Classify any given substance as ionic, molecular (nonpolar or polar), macromolecular, or metallic. Make qualitative predictions concerning its physical properties.

Example 11.6 and several problems at the end of the chapter illustrate this. You should be able to describe the typical properties associated with any one of these classes. Be able to account for relative melting and boiling points, solubilities, and electrical conductivities in terms of the units of structure and the forces between them. Table 11.5 neatly summarizes this information.

Consider the principles involved in the comparison of physical properties of *molecular* substances:

(a) Among substances of similar structure, melting point and boiling point ordinarily increase with molecular mass.

(b) Among substances of comparable molecular mass, those which are more polar usually have higher melting and boiling points.

(c) Hydrogen bonded species have unusually high melting and boiling points.

4. Write equations for the thermal decomposition of carbonates, hydroxides, and hydrates.

Equations 11.1–11.4 illustrate these reactions. So, too, does Example 11.5.

Additional descriptive material you should be familiar with includes:

(a) Be able to discuss the structures and properties of SiO_2 and several kinds of silicate minerals. See Section 11.3 in the text.

(b) Account for some of the properties of metals in terms of the electron sea model.

(c) Contrast the properties and structures of ice and liquid water.

5.   Determine ΔH for any given phase change.

Example 11.7 uses the data of Table 11.6, listing the thermal properties of water, to work two simple calculations. Both are of the sort you saw in Section 6.5 in the text on calorimetry. Review of Skill 5 in Chapter 6 of the guide may be useful here.

Note that in problems of this type it is assumed that no heat is lost to the surroundings. That is, the heat absorbed is numerically equal to that evolved. Again note that Example 11.7 requires the use of Equation 6.16:

Q = (specific heat) X m X Δt.

As noted in the discussion following Example 11.7 you may have to use Hess's Law to work some problems involving phase changes. This law was first described in Chapter 6 and will be used in still later chapters.

6.   **Given the phase diagram for a pure substance, decide what phase or phases may be present at a given temperature and pressure. Use the phase diagram to describe any phase change that occurs when T or P are changed.**

The phase diagram of water, Figure 11.13, is discussed in detail in the text. Some of the problems at the end of the chapter require the use of such a diagram, or even the construction of part of one from appropriate data. See the catalog of problems.

Consider the phase diagram for water. Each of the following points could also be made regarding any other phase diagram.
  — Within any area, such as that labeled liquid, only one phase is present.
  — Along any line, such as AB, two phases coexist in equilibrium.
  — At the triple point, A, all three phases are present.

_ _

Referring to the phase diagram for water, what phase(s) are present at point B? _____ What happens when the system described by point B is heated at constant pressure? _____

Point B is on the line separating liquid and gas phases. It represents a system containing both liquid and gaseous water in equilibrium at 25°C. (What is the equilibrium vapor pressure at this temperature?)

The addition of heat to such a system, held at constant pressure, causes evaporation of liquid. When all the liquid is converted to vapor, the temperature rises, as would be described by a horizontal line drawn through B and to the right of the liquid-vapor curve.

_ _

SELF-TEST

True or False

1. For any liquid, vapor pressure increases linearly with (is ( )
directly proportional to) the temperature.

2. Pure water cannot exist as a liquid below 0°C.   ( )

3. Generally, when a solid melts to form a liquid, the density ( )
decreases.

4. The heat of vaporization of a given substance is usually ( )
larger than the heat of fusion.

5. The physical properties of molecular substances are directly ( )
related to the strengths of the covalent bonds holding together the
molecules.

6. Low solubility in common solvents is characteristic of ( )
macromolecular and metallic substances.

Multiple Choice

7. When 1 mol liquid water is placed in a stoppered 200 cm^3 ( )
flask at 25°C, it eventually establishes a constant pressure P. If a
100 cm^3 flask were used instead, the final pressure would be
   (a) P/2   (b) P
   (c) 2P   (d) 1 atm

8. At any given temperature, which one of the following ( )
substances is likely to have the highest vapor pressure?
   (a) C_2H_6 (b) C_3H_8
   (c) C_4H_{10} (d) all will be the same

9. The heat of vaporization of NaCl is much larger than that of ( )
H_2O. This is a result of NaCl having
   (a) a larger formula mass
   (b) a larger density
   (c) a higher boiling point
   (d) stronger forces of attraction between units of structure

10. The triple point of I_2 is 0.12 atm and 115°C. This means ( )
that liquid iodine
   (a) is more dense than $I_2(s)$
   (b) cannot exist above 115°C

(c) cannot exist at 1 atm

(d) cannot have a vapor pressure less than 0.12 atm

11. Dry ice, $CO_2(s)$, is frequently used as a refrigerant. It ( )
undergoes sublimation at ordinary pressures. Energy is absorbed in
the process, overcoming
  (a) covalent bonds   (b) polar covalent bonds
  (c) dipole forces   (d) dispersion forces

12. Of the following interactions at the atomic-molecular level, ( )
the strongest is
  (a) dispersion force   (b) dipole force
  (c) hydrogen bond   (d) covalent bond

13. Which one of the following substances could be boiled ( )
without breaking hydrogen bonds?
  (a) H_2 (b) NH_3
  (c) H_2O_2 (d) none of these

14. Which of the following would you expect to be the best ( )
electrical conductor at room temperature?
  (a) H_2O (b) Al
  (c) SiO_2 (d) $CaCO_3$

15. The various types of glass you use in the laboratory are not ( )
likely to contain appreciable amounts of
  (a) boron   (b) carbon
  (c) oxygen   (d) silicon

16. Which class of substance would you expect to have the ( )
highest boiling points?
  (a) nonpolar molecular   (b) polar molecular
  (c) macromolecular   (d) organic

17. A given substance is a blue solid at 25°C. It decomposes at a ( )
temperature below 300°C to give a white solid and a second, volatile
product. The blue solid dissolves in H_2O to give a conducting
solution. Known to be one of the following, it is most likely
  (a) Cu   (b) CuF_2
  (c) $CuSO_4 \cdot 5 H_2O$ (d) a Cu–Hg amalgam

18. Small discrete molecules are expected to be found in solid ( )
  (a) $ZrSiO_4$ (b) Fe
  (c) CO_2 (d) KCl

19. Which of the following substances would be expected to ( )
have the highest boiling point?
  (a) Cl_2 (b) BrCl
  (c) HCl   (d) KCl

20. Silicates may exhibit ( )
    (a) covalent bonding (b) ionic bonding
    (c) dispersion forces (d) all of these

Problems

1. Write a balanced equation for any reaction occurring on heating
    (a) solid magnesium carbonate
    (b) $Ca(OH)_2$ (s)
    (c) water in an open beaker
    (d) ice, at a pressure below that of the triple point

2. What kind(s) of forces are overcome in melting each of the following substances? What are the units of structure which are being separated from each other?
    (a) $MgCl_2$ (b) CCl_4
    (c) HF (d) Fe

3. Classify each of the substances below as likely to be ionic, molecular, macromolecular, or metallic.

    A is a high-melting solid; a 50 g sample dissolves completely in 100 g water. _____

    B is a high-melting solid, insoluble in water, and an excellent conductor of heat. _____

    C is a gas at room temperature. _____

    D melts at 2400°C. Neither the solid nor the melt conducts an electric current. _____

4. One mole of water is introduced into an evacuated 100 cm^3 container at 82°C. At this temperature, it establishes its equilibrium vapor pressure of 0.50 atm.
    (a) What will be the pressure inside the container if the volume could be reduced to 50 cm^3?
    (b) What will be the pressure inside the 100 cm^3 container if the temperature is raised to 100°C?

5. The dependence of vapor pressure on the Kelvin temperature is given by the equation

$$Log_{10}(P_2/P_1) = \frac{\Delta H_{vap}}{2.30R}\left(\frac{T_2 - T_1}{T_2 T_1}\right)$$

If the vapor pressure of CCl_4(1) is 55% larger at 30°C than it is at 20°C, what do you calculate for the heat of vaporization? $R = 8.31 \text{ J/(mol·K)}$

SELF-TEST ANSWERS

1. **F** (You do not get a straight line when vapor pressure is plotted against T. Instead, vapor pressure rises exponentially.)
2. **F** (You need only increase the pressure to make ice melt below 0°C.)
3. **T** (Water is one of the few exceptions.)
4. **T** (The intermolecular forces are completely overcome in vaporization, only partially overcome in fusion.)
5. **F** (The intermolecular forces are more important.)
6. **T** (The dissolving of such substances requires overcoming the very strong forces between units of structure – e.g., the covalent bonds in a macromolecular structure.)
7. **b** (Vapor pressure depends only upon temperature.)
8. **a** (The smallest of these molecules should have the weakest dispersion forces.)
9. **d** (Ions are separated from each other; the strong ionic bond is overcome. In the case of H_2O, weaker forces between molecules are involved.)
10. **d** (The liquid-gas curve ends there.)
11. **d** (CO_2 is a nonpolar molecule. How do you know? See Skill 4 in Chapter 10 of this guide.)
12. **d** (The other three are seldom more than about 10% as strong.)
13. **a** (Boiling H_2 would involve overcoming only dispersion forces between molecules. Also, hydrogen bonding requires the presence of H–X, where X is F, O, or N.)
14. **b** (The only metal in the list.)
15. **b** (But will certainly contain the oxides of B and Si. See the discussion of silicates and glasses in the text.)
16. **c** (The very strong forces, covalent bonds, would have to be overcome.)
17. **c** (The decomposition suggests a polyatomic anion or hydrate. The volatile product is most likely CO_2 or H_2O. The conducting solution suggests an ionic solid.)
18. **c** (The other solids include two which are ionic, one metallic.)
19. **d** (The most polar, sufficiently so as to be ionic. A solid at 25°C.)
20. **d** (You might expect dispersion forces to hold together sheet-like structures as in talc and mica.)

Solutions to Problems

1. (a) $MgCO_3(s) \rightarrow MgO(s) + CO_2(g)$ (See Equation 11.2.)
   (b) $Ca(OH)_2(s) \rightarrow CaO(s) + H_2O(g)$ (See Equation 11.1.)
   (c) $H_2O(1) \rightarrow H_2O(g)$ (Evaporation or boiling.)
   (d) $H_2O(s) \rightarrow H_2O(g)$ (Sublimation.)

2. Forces overcome   Structural units separated
   (a) ionic bonds   ions
   (b) dispersion forces   molecules (nonpolar)
   (c) hydrogen bonds, as   molecules (polar)
       well as dipole and
       dispersion forces
   (d) metallic bonds   metal atoms (or ions)

3. *A* is ionic; *B*, metallic; *C*, molecular; *D*, macromolecular.

4. (a) 0.50 atm (Reducing the volume but maintaining constant T would result in condensation of some of the vapor.)
   (b) 1 atm (You should realize that the boiling point has been reached!)

5. $P_2 = 1.55 P_1$ or $P_2/P_1 = 1.55$
   $T_2 = 273 + 30 = 303$ K
   $T_1 = 293$ K

   $$Log_{10}(1.55) = \frac{\Delta H_{vap}}{2.30 (8.31)} \left(\frac{10}{303 \times 293} \right)$$

   $\Delta H_{vap} = 3.2 \times 10^4$ J/mol (See the similar text problem, 11.46.)

SELECTED READINGS

Intermolecular forces are also discussed in:

House, J.E., Weak Intermolecular Interactions, *Chemistry* (April 1972), pp. 13–15.

Liquids, their structure and properties, are discussed in:

Apfel, R.E., The Tensile Strength of Liquids, *Scientific American* (December 1972), pp. 58–71.
Bernal, J.D., The Structure of Liquids, *Scientific American* (August 1960), pp. 124–128.
Reid, R.C., Superheated Liquids, *American Scientist* (March-April 1976), pp. 146–156.
Turnbull, D., The Undercooling of Liquids, *Scientific American* (January 1965), pp. 38–46.

Solids, metallic, crystalline, and glassy, are discussed in the books and articles by Chaudhari, Fullman, Holden, Moore, and Sanderson listed in Chapter 9 and in:

Dandy, A.J., Chemistry of Cement, *Chemistry* (March 1978), pp. 13–16.

Darragh, P.J., Opals, *Scientific American* (April 1976), pp. 84–95.

Denio, A.A., Chemistry for Potters, *Journal of Chemical Education* (April 1980), pp. 272–275.

Epstein, A.J., Linear-Chain Conductors, *Scientific American* (October 1979), pp. 52–61.

Greene, C.H., Glass, *Scientific American* (January 1961), pp. 92–105.

Kolb, D., The Chemistry of Glass, *Journal of Chemical Education* (September 1979), pp. 604–608.

McQueen, H.J., The Deformation of Metals at High Temperatures, *Scientific American* (April 1975), pp. 116–125.

Various theories of the metallic bond are considered in the books on bonding listed in Chapter 10.

Water, that most important liquid and solid:

Frank, H.S., The Structure of Ordinary Water, *Science* (August 14, 1970), pp. 635–641.

Runnels, L.K., Ice, *Scientific American* (December 1966), pp. 118–126.

12

THE NONMETALS AND THEIR
COMPOUNDS (GROUPS 4–8)

QUESTIONS TO GUIDE YOUR STUDY

1. How do the nonmetals occur in nature? For each element itself, what is the formula, the physical state, and atomic-molecular structure at 25°C and 1 atm?

2. What properties, physical and chemical, are characteristic of the nonmetals? The metalloids?

3. What are the expected trends in atomic and physical properties among the elements? What explanations can be given in terms of electronic structure? What correlations can be made with position in the Periodic Table?

4. What reactions do the nonmetals take part in among themselves? What are some of the properties and reactions of their simpler compounds?

5. How do you account for the common occurrence of more than one oxide of a given nonmetal? What conditions favor the formation of one instead of another?

6. How do you account for the vast number and variety of carbon compounds? How do isomers differ from one another? Can you explain their differences in terms of molecular structure and intermolecular forces?

7. How would a useful compound, such as the anesthetic N_2O, be prepared in the laboratory? How is it prepared commercially?

8. What are some of the uses of other nonmetals and their compounds? Of the metalloids, such as silicon and germanium?

9. We have constructed atoms by filling atomic orbitals. Is there a way we can construct molecules, such as O_2, by filling **molecular** orbitals (rather than **hybrid** orbitals)?

YOU WILL NEED TO KNOW

Like Chapter 9, this chapter is mostly a review and illustration of principles that you have already seen in detail. A review of the material in

Sections 5.1 and 5.2 would serve as a useful perspective for this chapter. More specifically useful as background is the following.

Concepts

　　1.　How to write valence electron configurations for the elements of Groups 4 to 8. See Skill 7 in Chapter 8 of this guide.
　　2.　How to draw Lewis structures for molecules and polyatomic ions. How to use a Lewis structure to predict geometry and polarity. How to decide the kind and number of bonds (sigma and pi) in a species. See the skills of Chapter 10 in the guide.
　　3.　What is meant by hybridization. See Section 10.5 in the text.
　　4.　How to predict the kinds of intermolecular forces present in a given molecular substance. How to make qualitative predictions comparing physical properties of molecular substances. Chapter 11.

Math

　　Essentially a nonmathematical chapter.
　　1.　How to write balanced equations for chemical reactions – Chapter 4.

BASIC SKILLS

　　Many of the skills used in this chapter have already been discussed. Among these are the following:
　　– *Given the formula of a molecular species or a polyatomic ion, write the Lewis structure.* Examples 12.1–12.4 as well as 12.6 and 12.7 require this. If you need a review, see Skill 2 in Chapter 10.
　　– *Given the Lewis structure of a species, predict a bond angle or the overall geometry.* Examples 12.3 and 12.6 illustrate this. The prediction is based on the principle of electron-pair repulsion and was seen in Skill 3 of Chapter 10.
　　– *Given the geometry of a species, predict whether or not it is polar.* This skill (Skill 4 in Chapter 10) is illustrated by Example 12.6.
　　– *Given the Lewis structure of a species, indicate the kind and number of bonds (sigma and pi).* This is Skill 6 in Chapter 10 and is shown again here in Example 12.7.
　　– *Make qualitative comparisons of physical properties of molecular substances.* Example 12.5 illustrates the use of this and the related skill, Skills 2 and 3 in Chapter 11.

Skills new to this chapter include the following:

1. **Given or having derived the Lewis structure for a molecule or polyatomic ion, write plausible resonance structures.**

You should expect resonance for any species for which you can write more than one reasonable Lewis structure using a given skeleton (bonding sequence) of atoms. In particular, look for it if your Lewis structure contains multiple bonds.

Resonance constitutes, in essence, an admission that we cannot represent a particular species by one simple Lewis structure. The species has only one structure; the problem is to find a satisfactory way of describing it! So we imagine it to look like a mixture, or hybrid, of these resonance structures.

Examples 12.1 and 12.6 ask for resonance structures. So do some of the problems at the end of the chapter. See the catalog of problems.

2. **Given the molecular formula of an alkane, draw the structural formula for each possible isomer.**

Example 12.4 illustrates this skill. Consider the exercise that directly follows that example. There you are asked for the isomers with the molecular formula C_6H_{14}. In any such problem it is probably best to use a systematic approach. Here you might start by drawing the isomer with a 6-carbon chain, then all the isomers with 5-carbon chains, and so on. The most common mistake is to draw too many formulas. You must realize, for instance, that the two members of each of the following pairs are actually identical:

$$
\begin{array}{c}
 C \\
 | \\
C-C-C-C-C-C \quad and \quad C-C-C-C-C
\end{array}
$$

(both have a 6-carbon chain)

$$
\begin{array}{c}
C-C-C-C-C \qquad and \quad C-C-C-C-C \\
| | \\
C C
\end{array}
$$

(both have a 1-carbon atom branch on the second atom in the chain)

A molecular model kit is very helpful in constructing isomers. If you make models of such pairs as those above, it will be immediately obvious that they are identical.

The existence of isomers (and there are others in addition to structural) is one reason for the abundance and variety of carbon compounds. The formation of strong carbon-carbon bonds is another.

As for other molecular substances, you should be able to qualitatively compare the physical properties of isomers. Example 12.5 does this, comparing the boiling points of the isomers with the formula C_5H_{12}.

3. Given the formula of an expanded octet species, state the number of electron pairs around the central atom and give the hybridization. Describe the geometry of the species.

Examples 12.8 and 12.9 illustrate these skills. Figure 12.10 shows the geometries usually observed for such species. Note that although there are seven new ways of describing geometry (octahedron, square pyramid, etc.), there are only two new hybrid orbitals to learn: sp^3d (5 pairs of electrons) and sp^3d^2 (6 pairs).

Describe the geometry of BrF_3; of BrF_4^-. _____
First you need to draw the structures, then predict the geometry. Counting the valence electrons around the central atom (Br): 1 from each F and 7 from Br = 10. Using Figure 12.10, we see that 5 pairs, 3 of them bonding, give a T-shaped species. (Also, BrF_3 is analogous to ClF_3. Both involve atoms of elements from the same group and similar formulas. You should expect structures to be similar, too.)
For BrF_4^-, there is 1 electron due to the charge, plus 1 more from the additional fluorine, for a total of 12. There are 6 pairs, of which 4 are bonding. The geometry expected, again using Figure 12.10, is square planar, like that of XeF_4, which has the same total number of valence electrons.

----

4. Write the molecular orbital diagram for a simple diatomic species.

The principles here are entirely analogous to those used in Chapter 8 to write atomic orbital diagrams for isolated atoms. Each molecular orbital, like each atomic orbital, can hold two electrons of opposite spin. When two orbitals of equal energy are available, electrons enter singly with parallel spins. Note the structures of B_2 and O_2 in Table 12.4. Unfortunately, to work out molecular orbital diagrams, you need to use new symbols and a new order of filling orbitals. The latter can be deduced from Table 12.4 as being: $\sigma_{1s}^b \sigma_{1s}^* \sigma_{2s}^b \sigma_{2s}^* \pi_{2p}^b \pi_{2p}^b \sigma_{2p}^b \pi_{2p}^*$ .... A few of the problems require this skill. Several diagrams are worked out in the body of the text. See Section 12.5.

5. **Write balanced equations for some of the reactions of a nonmetal with H_2 ; O_2 ; or a halogen.**

This chapter contains a considerable amount of new descriptive material. Some of it is organized around such reactions as are described by Equations 12.1–12.12. Besides writing such equations (and even working problems based on them), you should be able to discuss the conditions under which these reactions occur. Example: what conditions favor the formation of a higher oxide (say CO_2 or NO_2) instead of a lower oxide (CO or NO)? See the catalog of problems for additional illustrations.

SELF-TEST

True or False

1. All the Group 4 elements are solids at 25°C, 1 atm.   (   )

2. For molecular substances of similar structure, boiling point   (   )
increases with molecular mass.

3. Diamond and graphite are isomers of carbon.   (   )

4. A saturated hydrocarbon is one in which each carbon atom   (   )
is bonded to four other carbon atoms.

5. Two resonance structures differ from each other in the   (   )
sequence in which the atoms are bonded.

Multiple Choice

6. Which one of the following nonmetals does *not* show   (   )
allotropy?
    (a) O   (b) N   (c) C   (d) S

7. The most reactive of the halogens is   (   )
    (a) F_2 (b) Cl_2
    (c) Br_2 (d) I_2

8. Compounds of bromine and another Group 7 element (X)   (   )
are expected to have the formula
    (a) BrX   (b) BrX_3
    (c) BrX_5 (d) any of these

9. Any Group 6 atom is likely to form an ion with a charge of ( )
    (a) +6   (b) +2
    (c) -2   (d) -6

10. The fact that all bond distances in SO_3 are the same is ( )
accounted for by
    (a) the idea of resonance   (b) electron-pair repulsion
    (c) the octet rule   (d) the existence of isomers

11. Which would be described in terms of an "expanded" octet? ( )
    (a) SCl_2 (b) SCl_4
    (c) neither of these   (d) both of these

12. You would expect SF_6 to have _____ electron pairs ( )
around the central atom.
    (a) 12   (b) 6
    (c) 4   (d) some other number

13. What geometry is associated with $sp^3 d^2$ hybridization?   ( )
    (a) octahedral   (b) square pyramidal
    (c) square planar   (d) any one of these is
                                          possible

14. The fact that $SbCl_5$ has the shape of a trigonal bipyramid is ( )
explained in terms of _____ hybridization of Sb.
    (a) sp   (b) sp^2
    (c) sp^3 (d) $sp^3 d$

15. The description of the electronic structure of O_2 that best ( )
accounts for both the bond energy and magnetic properties is given
by
    (a) :Ö=Ö: (b) :Ö-Ö:
    (c) a resonance hybrid of structures (a) and (b)
    (d) molecular orbital theory

16. Which one of the following resonance structures of CO_2 is ( )
the least plausible?
    (a) :Ö-C≡O: (b) :O≡C-Ö:

    (c) :Ö=C=Ö: (d) :Ö-C-Ö:

17. Covalent bonds are broken during at least some phase ( )
changes in
    (a) sulfur   (b) graphite
    (c) red phosphorus   (d) all of these

18. Ammonia boils at a higher temperature than does phosphine ( )
(PH_3). This is primarily due to NH_3
    (a) having a smaller mass
    (b) being polar
    (c) having larger bond angles
    (d) exhibiting hydrogen bonds

19. If you wanted to prepare the lower of the two common ( )
oxides of sulfur, you would burn sulfur in
    (a) limited air
    (b) excess air
    (c) limited air and cool in the absence of air
    (d) pure oxygen

20. Which of the following could function as a semiconductor ( )
in which electric charge is carried by electrons?
    (a) very pure Si
    (b) very pure Ge
    (c) very pure Si to which has been added a trace of B
        (Group 3)
    (d) very pure Si to which has been added a trace of As
        (Group 5)

Problems

1. Draw structural formulas for the isomers with molecular formula C_6H_{14}. Which one should have the highest boiling point?

2. Draw resonance structures and predict the bond angles and the hybridization of N in:
    (a) NO_3^- (b) NO_2^+

3. Write balanced equations for any reactions occurring when
    (a) carbon is heated to high T in a limited amount of air
    (b) hot nitric oxide is allowed to cool in air
    (c) ammonium nitrate is carefully heated

4. Complete the following table:

Formula	Molecular mass	No. of electron pairs	Hybridization	Geometry
SeF_6	193	_____	_____	_____
PCl_5	208	_____	_____	_____

5. Taking the order of filling molecular orbitals to be:

$$\sigma_{2s}^{b} < \sigma_{2s}^{*} < \pi_{2p}^{b} = \pi_{2p}^{b} < \sigma_{2p}^{b} < \pi_{2p}^{*} = \pi_{2p}^{*}$$

arrange O_2^{+}, O_2, O_2^{-}, and O_2^{2-} in order of increasing bond energy.

SELF-TEST ANSWERS

1. **T** (The first three are macromolecular; Sn and Pb are common metals. See Table 12.1.)
2. **T** (Dispersion forces increase in strength as molecular mass increases. Chapter 11.)
3. **F** (They are allotropic forms.)
4. **F** (Four atoms, yes, but not four carbon atoms. Some of the carbons must be bonded to hydrogen. Otherwise, the description fits diamond, not a hydrocarbon.)
5. **F** (Resonance structures differ only in the placement of valence electrons. Isomers differ in bonding sequence.)
6. **b**
7. **a** (A question that was asked in Chapter 5.)
8. **d** (The first formula is analogous to Br_2, an octet structure. The others are expanded octet structures: BrX_3 would have 5 pairs of valence electrons, BrX_5 would have 6 pairs.)
9. **c** (To give an octet to these highly electronegative atoms.)
10. **a** (All bonds in SO_3 are intermediate between a single and a double bond, as suggested by a hybrid of resonance structures.)
11. **b** (A simple octet structure can be written for SCl_2. SCl_4 has 6 plus 4 valence electrons around the central sulfur atom.)
12. **b** (Six electrons from S + one from each of six F = 12 electrons, or 6 pairs. The geometry would be octahedral.)
13. **a** (As predicted by electron-pair repulsion. See Figure 12.10.)
14. **d** (Here, as for the preceding question, see the discussion in Section 12.4.)
15. **d** (One of the few reasons for considering molecular orbital theory in an introduction to chemistry.)
16. **d** (Note the absence of an octet on carbon.)
17. **d** (All of these are macromolecular.)
18. **d** (Without this, NH_3 should have the lower boiling point. The effects of intermolecular forces on physical properties was considered in detail in Chapter 11.)

19. **c** (Cooled in the presence of air would give the higher oxide, SO_3 instead of SO_2.)

20. **d** (As carries an additional valence electron, giving an n-type semiconductor.)

Solutions to Problems

1. The skeletons would look like:

    C–C–C–C–C–C   C–C–C–C–C   C–C–C–C
    $\qquad\qquad\qquad\qquad\qquad$ | $\qquad\qquad\qquad\quad$ | |
    $\qquad\qquad\qquad\qquad\qquad$ C $\qquad\qquad\qquad\quad$ C C

    $\qquad\qquad\qquad\qquad\qquad\qquad\qquad\qquad\quad$ C
    $\qquad\qquad\qquad\qquad\qquad\qquad\qquad\qquad\quad$ |
    $\qquad\qquad$ C–C–C–C–C $\qquad\quad$ C–C–C–C
    $\qquad\qquad\qquad\quad$ | $\qquad\qquad\qquad\qquad\quad$ |
    $\qquad\qquad\qquad\quad$ C $\qquad\qquad\qquad\qquad\quad$ C

(To complete the structures, add –H to give octets to all carbon atoms.)

$\qquad$ The highest boiling would be the least compact, the 6-C chain given first.

2. Except for the resonance structures, this question belongs in Chapter 10!

(a) NO_3^- $\qquad$:Ö–N=Ö: ⟷ :Ö=N–Ö: ⟷ :Ö–N–Ö:
$\qquad\qquad\qquad\qquad\quad$ | $\qquad\qquad\qquad\quad$ | $\qquad\qquad\qquad\quad$ ‖
$\qquad\qquad\qquad\qquad\quad$:O: $\qquad\qquad\qquad\quad$:O: $\qquad\qquad\qquad\quad$:O:

Bond angles = 120° $\qquad$ Hybrids: sp^2

(b) NO_2^+ $\qquad$:Ö=N=Ö: ⟷ :Ö–N≡O: ⟷ :O≡N–Ö:

Bond angles = 180° $\qquad$ Hybrids: sp

3. (a) $2\ C(s) + O_2(g) \rightarrow 2\ CO(g)$ $\qquad$ (Equation 12.9.)
   (b) $2\ NO(g) + O_2(g) \rightarrow 2\ NO_2(g)$ $\qquad$ (Equation 12.7.)
   (c) $NH_4NO_3(s) \rightarrow N_2O(g) + 2\ H_2O(g)$ $\quad$ (Equation 12.8.)

4. 6 pairs $\qquad sp^3d^2$ hybrids $\qquad$ octahedral
   5 pairs $\qquad sp^3d$ hybrids $\qquad$ trigonal bipyramidal

5. In each of these species all orbitals through σ_{2p}^b are filled. The remainder of the molecular orbital diagram looks like:

	π_{2p}^*	π_{2p}^*	
O_2^+	(↑)	( )	with 2.5 electron pair bonds, total
O_2	(↑)	(↑)	with 2.0 electron pair bonds, total
O_2^-	(↑↓)	(↑)	with 1.5 electron pair bonds, total
O_2^{2-}	(↑↓)	(↑↓)	with 1.0 electron pair bonds, total

So, in order of increasing number of bonds, read this column bottom to top. This is the same order as that of increasing bond energy and decreasing bond distance. Problem 12.21 in the text is similar.

SELECTED READINGS

Chemical bonds are discussed in the books listed in Chapter 10 and:

Moody, G.J., A Decade of Xenon Chemistry, *Journal of Chemical Education* (October 1974), pp. 628–630.

Selig, H., The Chemistry of the Noble Gases, *Scientific American* (May 1964), pp. 66–77.

Ward, R., Would Mendeleev Have Predicted the Existence of XeF_4? *Journal of Chemical Education* (May 1963), pp. 277–279.

Metalloids, nonmetals, and other materials are discussed in:

Adler, D., Amorphous-Semiconductor Devices, *Scientific American* (May 1977), pp. 36–48.

Allcock, H.R., Inorganic Polymers, *Scientific American* (March 1974), pp. 66–74.

Bundy, F.P., Superhard Materials, *Scientific American* (August 1974), pp. 62–70.

Chedd, G., *Half-Way Elements: The Technology of Metalloids,* Garden City, N.J., Doubleday, 1969.

Dence, J.B., Covalent Carbon-Metal(loid) Compounds, *Chemistry* (January 1973), pp. 6–13.

Derjaguin, B.V., The Synthesis of Diamond at Low Pressure, *Scientific American* (November 1975), pp. 102–109.

Johnston, W.D., The Prospects for Photovoltaic Conversion. *American Scientist* (November-December 1977), pp. 729–736.

Materials, *Scientific American* (September 1967).

Rochow, E.G., *The Metalloids,* Boston, D.C. Heath, 1966.

Reactions of industrial importance are considered in:

Cook, G.A., *Survey of Modern Industrial Chemistry,* Ann Arbor, Mich., Ann Arbor Science, 1975.

13

CHEMICAL EQUILIBRIA IN GASEOUS SYSTEMS

QUESTIONS TO GUIDE YOUR STUDY

1. What is a dynamic equilibrium?
2. What is an equilibrium vapor pressure? What conditions determine its value? How rapidly is such a pressure established?
3. How would you show experimentally that a given system is in a state of equilibrium?
4. If you could watch the molecules in an equilibrium system, what would you expect to see?
5. How can you predict the conditions under which equilibrium may exist? For example, at what temperature and pressure are ice and water in equilibrium? Or hydrogen, oxygen, and water?
6. Can you predict the effect of changes in conditions on a system already at equilibrium?
7. Is there a general approach to describing equilibrium systems? (You have already seen phase equilibria treated. Later chapters deal with several types of equilibria in water solution.)
8. Can you describe a common example of a reaction that only partially converts reactants to products?
9. If many reactions do not go to completion, what does it mean about how you are to interpret chemical equations for such reactions?
10. How are equilibrium calculations useful?

YOU WILL NEED TO KNOW

Concepts

1. How to interpret the sign of ΔH. See Chapter 6.
2. The general nature of an equilibrium system, as described in Chapter 11, Section 11.1.

Math

1. How to work problems in stoichiometry (including concentrations). See Chapter 4.

2. How to take square roots on your calculator. Examples:

$(4.0)^{1/2} = 2.0$

$(4.0 \times 10^{-2})^{1/2} = 2.0 \times 10^{-1} = 0.20$

$(4.0 \times 10^{-3})^{1/2} = 6.3 \times 10^{-2} = 0.063$

3. How to solve quadratic equations (equations involving x^2 terms). Often, this can be done by taking the square roots of both sides. For example, to solve:

$$\frac{x^2}{(1 - 2x)^2} = 4.0 \times 10^{-3}$$

take the square root of both sides:

$$\frac{x}{1 - 2x} = 0.063$$

$$x = 0.063 - 0.126 x; \quad 1.126 x = 0.063; \quad x = 0.056$$

If the terms involving x do not form a perfect square, as in

$$\frac{x^2}{1 - x} = 0.20$$

a more general approach is required. This involves using the quadratic formula (Example 13.6). Here the equation is first rearranged to get it in the form:

$$ax^2 + bx + c = 0$$

where a, b, and c are numbers. Then apply the formula:

$$x = \frac{-b \pm \sqrt{b^2 - 4ac}}{2a}$$

For the equation above:

$$x^2 = 0.20 - 0.20x$$

$$x^2 + 0.20x - 0.20 = 0; \quad a = 1, b = 0.20, c = -0.20$$

$$x = \frac{-0.20 \pm \sqrt{0.04 + 0.80}}{2} = \frac{-0.20 \pm \sqrt{0.84}}{2}$$

$$x = \frac{-0.20 \pm 0.92}{2} = -0.56 \text{ or } 0.36$$

In chemical systems such as the ones discussed in this chapter, only one of the two possible solutions for x will make sense.

BASIC SKILLS

1. **Given the balanced equation for a reaction involving one or more gases, write the corresponding expression for the equilibrium constant, K_c.**

The principles involved here are discussed in Section 13.3 of the text. Note particularly that terms for pure solids and liquids do not appear in the expression for K_c.

- -

What is the expression for K_c for:

$$4\ NH_3(g) + 7\ O_2(g) \rightleftharpoons 4\ NO_2(g) + 6\ H_2O(l)?\ \underline{\hspace{1.5cm}}$$

We note that the term for the gaseous product, NO_2, appears in the numerator, while those for the reactants, NH_3 and O_2, are in the denominator. No term for H_2O is included since it is a liquid. The exponents of the concentration terms are the coefficients in the balanced equation.

$$K_c = \frac{[NO_2]^4}{[NH_3]^4[O_2]^7}$$

- -

See Example 13.3 and the catalog of problems in the text for further illustration of this skill.

2. **Given the equation for a reaction, calculate K_c knowing:**

a. **the equilibrium concentrations of all species.**

Example 13.2 illustrates this skill, as does the following example.

- -

For the reaction: $2\ NO(g) + O_2(g) \rightleftharpoons 2\ NO_2(g)$, the equilibrium concentrations of NO_2, NO, and O_2 are 1.0×10^{-2}, 1.0×10^{-3}, and 0.50×10^{-1} M, respectively. What is K_c? $\underline{\hspace{1.5cm}}$

All that is required here is to set up the expression for K_c and substitute the concentrations given:

$$K_c = \frac{[NO_2]^2}{[NO]^2\,[O_2]} = \frac{(1.0 \times 10^{-2})^2}{(1.0 \times 10^{-3})^2\,(0.50 \times 10^{-1})} = 2.0 \times 10^3$$

----

b. **the original concentrations of all species and the equilibrium concentration of one.**

As in working most of the problems involving equilibria, it may be helpful to set up a table of concentrations as in the following example.

For the reaction: $2\,HI(g) \rightleftharpoons H_2(g) + I_2(g)$, if we start with pure HI at a concentration of 1.20 M, that of H_2 at equilibrium is 0.20 M. What is K_c for the reaction? _____

To work this problem, you must recall the significance of the coefficients of a balanced equation: they give the mole ratios between reactants and products. Since H_2 and I_2 both have coefficients of 1, it follows that when 0.20 M of H_2 is produced, an equal number, or 0.20 M, of I_2 must be formed simultaneously. Since 2 mol of HI are required to give 1 mol of H_2, 2(0.20) = 0.40 M HI must be consumed to produce 0.20 M H_2. Finally, 1.20 – 0.40 = 0.80 M of HI must be left. Summarizing this reasoning in the form of a table:

	orig. conc. (M)	change (M)	equil. conc. (M)
HI	1.20	-0.40	0.80
H_2	0.00	+0.20	0.20
I_2	0.00	+0.20	0.20

With this information, K_c is readily calculated:

$$K_c = \frac{[H_2]\,[I_2]}{[HI]^2} = \frac{(0.20)\,(0.20)}{(0.80)^2} = 0.062$$

Example 13.1 illustrates part of the solution we have just worked through. Given the equation for a reaction, the original concentrations, and one equilibrium concentration, you should be able to determine all the other equilibrium concentrations. All changes in concentrations *must* be related through the use of the coefficients of the balanced equation for the reaction.

3. **Given the value of K_c.**

a. **and all but one equilibrium concentration, calculate the remaining equilibrium concentration.**

This is illustrated by Example 13.4.

b. **and all the original concentrations, calculate the equilibrium concentrations.**

This is probably the most useful application of the equilibrium constant expression. It is also the one that causes the most difficulties. Work through Examples 13.5 and 13.6 carefully. It may be worth cautioning against some of the mistakes commonly made in analyzing problems of this type.

– Do not confuse original and equilibrium concentrations. Note in Example 13.5 that since there is originally no product NO, some of it must form in order to reach equilibrium. In so doing, some of each reactant must be consumed.

– Be sure you relate properly the changes in reactant and product concentrations. Again referring to Example 13.5, students occasionally take the change in NO concentration to be x rather than 2x, forgetting that the coefficients of the balanced equation require a 2:1 mol ratio of NO to N_2.

Calculations of this sort are illustrated more simply by Example 13.5 (no quadratic formula required, provided you recognize the perfect square). Example 13.6 requires the use of the quadratic formula.

4. **Given the value of K_c, predict the direction in which a chemical system will move to reach equilibrium and determine the equilibrium concentrations.**

Example 13.7 applies this skill to a system originally at equilibrium. The system is disturbed by the addition of a reactant and you are asked to determine the new equilibrium concentrations.

5. **Using Le Chatelier's Principle, predict the effect of**
   a. **adding (or removing) a reactant or product,**
   b. **changing the volume, or**
   c. **changing the temperature**
upon the composition of an equilibrium system.

Qualitative predictions of this type are called for in Example 13.8. Example 13.7, as noted in Skill 4 above, calls for a quantitative prediction.

SELF-TEST

True or False

1. Higher temperatures would favor the production of more ( )
product in the system: benzene(l) $\rightleftharpoons$ benzene(g).

2. The expression for K_c always shows all gaseous species, but ( )
not pure solid or liquid species.

3. The value of K_c for a given reaction depends on the initial ( )
concentrations of reactants.

4. The larger the value of K_c, the more extensive the ( )
conversion of reactants to products.

5. The value of K_c is expected to increase with temperature ( )
for any reaction that has a negative value of ΔH.

6. Two students study the equilibrium between hydrogen, ( )
oxygen, and liquid water. One student calculates K_c using the
equation $2 H_2(g) + O_2(g) \rightleftharpoons 2 H_2O(l)$. The other student calculates
K_c using the equation $H_2(g) + \frac{1}{2} O_2(g) \rightleftharpoons H_2O(l)$. Their values for K_c
should be the same.

Multiple Choice

7. The expression for K_c for the equilibrium $C(s) + CO_2(g) \rightleftharpoons$ ( )
$2 CO(g)$ is:

(a) $\dfrac{2 [CO]}{[C] [CO_2]}$

(b) $\dfrac{[CO]^2}{[C] [CO_2]}$

(c) $\dfrac{[CO]^2}{[CO_2]}$

(d) $\dfrac{[CO]}{[CO_2]^2}$

8. Approximately equal amounts of reactants are mixed in a ( )
suitable container. Given sufficient time, the reactants may be
converted almost entirely to products if
    (a) K_c is much less than one
    (b) K_c is much larger than one
    (c) $K_c = 1$
    (d) $K_c = 0$

9. At a certain temperature, $K_c = 1$ for the reaction $2 HCl(g) \rightleftharpoons$ ( )
$H_2(g) + Cl_2(g)$. In this system at equilibrium, then, one can be sure
that
  (a) $[HCl] = [H_2] = [Cl_2] = 1$
  (b) $[H_2] = [Cl_2]$
  (c) $[HCl] = 2 \times [H_2]$
  (d) $\dfrac{[H_2][Cl_2]}{[HCl]^2} = 1$

10. One mole of HI(g) and 1 mol $H_2(g)$ are placed in an ( )
evacuated container at $100°C$ and allowed to come to equilibrium.
The reaction is $2 HI(g) \rightleftharpoons H_2(g) + I_2(g)$. As equilibrium is
approached, you can be sure that
  (a) I_2 concentration will increase
  (b) H_2 concentration will increase
  (c) HI concentration will decrease
  (d) all the above

11. What would you predict to be the conditions that would ( )
favor maximum conversion of noxious nitric oxide and carbon
monoxide?

$$NO(g) + CO(g) \rightarrow \tfrac{1}{2} N_2(g) + CO_2(g), \Delta H = -374 \text{ kJ}$$

  (a) low T, high P   (b) high T, high P
  (c) low T, low P   (d) high T, low P

12. The position of equilibrium would not be appreciably ( )
affected by changes in container volume for
  (a) $H_2(g) + I_2(s) \rightleftharpoons 2 HI(g)$
  (b) $N_2(g) + O_2(g) \rightleftharpoons 2 NO(g)$
  (c) $N_2(g) + 3 H_2(g) \rightleftharpoons 2 NH_3(g)$
  (d) $H_2O_2(l) \rightleftharpoons H_2O(l) + \tfrac{1}{2} O_2(g)$

13. Which of the following changes will invariably increase the ( )
yield of products at equilibrium?
  (a) an increase in temperature
  (b) an increase in pressure
  (c) addition of a catalyst
  (d) increasing original reactant concentrations

14. If we start with one mole of N_2, three moles of H_2, and two ( )
moles of NH_3 in a container at $500°C$, at equilibrium
  (a) the number of moles of N_2, H_2, and NH_3 will be in the
       ratio $1:3:2$

(b) the number of moles of N_2 and H_2 will be in the ratio 1:3

(c) the number of moles of N_2 will be one

(d) the total number of moles will be six

15. If a system contains SO_2, O_2, and SO_3 gases at equilibrium,   ( )

$$SO_2(g) + \tfrac{1}{2} O_2(g) \rightleftharpoons SO_3(g)$$

an increase in the partial pressure of SO_3 brought about by the addition of more SO_3 to the system will result in

(a) a reaction in which some SO_3 is formed

(b) a reaction in which all of the added SO_3 is consumed

(c) a reaction in which some of the added SO_3 is consumed

(d) no reaction

16. It is often the case that K_c is very small for a reaction under   ( )
almost all conditions. Yet the reaction may be used to produce significant amounts of products. How can this be?

(a) an alchemist is employed to increase the yield

(b) the reaction is carried out at very high temperatures

(c) an alternate series of reactions, giving the same net result, is used

(d) product is removed from the system as it is formed

17. For the reaction $CaO(s) + CO_2(g) \rightleftharpoons CaCO_3(s)$, we find that   ( )
$K_c = 277$ at 800°C. What is $[CO_2]$?

(a) 277

(b) (1/277)

(c) $(277)^{\frac{1}{2}}$

(d) it depends on the amounts of CaO and CO_2

18. For the reaction $2 SO_3(g) \rightleftharpoons 2 SO_2(g) + O_2(g)$, $K_c = 32$. If   ( )
$[SO_3] = [O_2] = 2.0$ M, then $[SO_2]$ is

(a) 0.031   (b) 0.25

(c) 5.7   (d) 8.0

19. The extent of reaction is negligible in the case of   ( )

(a) $K_c = 10^{10}$ (b) $K_c = 1$

(c) $K_c = 10^{-1}$ (d) $K_c = 10^{-10}$

20. For the reaction $H_2(g) + Br_2(g) \rightleftharpoons 2 HBr(g)$, $K_c = 4.0 \times$ ( )
10^{-2}. For the reaction $2 HBr(g) \rightleftharpoons H_2(g) + Br_2(g)$, K_c must be

(a) 4.0×10^{-2} (b) 2.0×10^{-1}

(c) 5.0   (d) 25

Problems

1. Consider the reaction $2 SO_3(g) \rightleftharpoons 2 SO_2(g) + O_2(g)$.
    (a) Write the expression for K_c.
    (b) Calculate K_c if in an equilibrium system it is found that the concentrations of SO_2, O_2, and SO_3 are 1.6×10^{-2}, 1.2×10^{-1}, and 3.2×10^{-2} M, respectively.

2. For the reaction $2 HI(g) \rightleftharpoons H_2(g) + I_2(g)$, $K_c = 0.010$ at 500°C. Calculate the concentrations of all species at equilibrium, starting with 0.30 mol HI in a 5.0 ℓ (5.0 dm^3) container.

3. Consider the equilibrium $SnO_2(s) + 2 H_2(g) \rightleftharpoons Sn(s) + 2 H_2O(g)$.
    (a) Write the expression for K_c.
    (b) At a certain temperature, $K_c = 0.25$. At equilibrium in a container with a volume of 2.0 ℓ (2.0 dm^3), there are 1.5 mol Sn, 1.0 mol H_2, and 1.2 mol SnO_2. What is the concentration of H_2O?

4. Consider the system $2 N_2O(g) \rightleftharpoons 2 N_2(g) + O_2(g)$, $\Delta H = + 163$ kJ. In which direction will an equilibrium system move to reestablish equilibrium if
    (a) N_2O is added?
    (b) O_2 is removed?
    (c) The volume is increased?
    (d) The temperature is raised?

5. Consider the reaction $2HI(g) \rightleftharpoons H_2(g) + I_2(g)$. At 1000°C, 25% of a sample of HI is found to decompose to the elements at equilibrium. Calculate K_c.

SELF-TEST ANSWERS

1. **T** (This kind of prediction was made and explained in Chapter 11. Here, you see it predicted by Le Chatelier's Principle: vaporization is an endothermic process.)
2. **T** (See the discussion following Example 13.2)
3. **F** (K_c for a given reaction depends only on T.)
4. **T** (See Section 13.4.)

5. **F** (Decrease. You can predict this from Le Chatelier's Principle. See Section 13.6 in the text.)

6. **F** (The value of K_c obtained by the second student should be the square root of that obtained by the first student; compare the two expressions.)

7. **c** (Do not include the solid.)

8. **b** (See Question 4.)

9. **d** (The composition could be anything, provided this ratio equals K_c.)

10. **d** (There must be some I_2 for equilibrium to exist. And if I_2 forms, HI must be consumed and some additional H_2 formed as well.)

11. **a** (Low T favors the exothermic reaction. High P favors fewer moles of gas.)

12. **b** (There is no change in the number of moles of gas.)

13. **d** (Yield in the sense of total amount of products. What about the percentage of reactants converted to products?)

14. **b** This is the ratio of moles we started with and, more important generally, the ratio of moles that react — as given by the coefficients in the equation for the reaction.)

15. **c** (Apply Le Chatelier's Principle. The number of moles of SO_3 has been temporarily increased. Note that P has not increased for the other components.)

16. **d** (Removal of product shifts the equilibrium so as to form more product.)

17. **b** ($K_c = 1/[CO_2]$.)

18. **d** (Set up the expression for K_c and substitute the concentrations of SO_3 and O_2.)

19. **d** (See Section 13.4.)

20. **d** (Write out the expressions for two different K_c's and see how they relate to each other.)

Solutions to Problems

1. (a) $K_c = [SO_2]^2 [O_2]/[SO_3]^2$

   (b) Substituting these equilibrium concentrations in the above expression: $K_c = (1.6 \times 10^{-2})^2 (1.2 \times 10^{-1})/(3.2 \times 10^{-2})^2$
   $$= 3.0 \times 10^{-2}$$

2. $K_c = 0.010 = [H_2][I_2]/[HI]^2$

   Setting up a table of concentrations:

	original	change	final
HI	0.30/5.0 = 0.060	-2x	0.060 - 2x
H_2	0.000	+x	x
I_2	0.000	+x	x

and substituting in the expression for K_c:
$$0.010 = (x)(x)/(0.060 - 2x)^2$$
taking the square root of both sides:
$$0.10 = x/(0.060 - 2x)$$
$$x = 0.0050 = [H_2] = [I_2]$$
$$[HI] = 0.060 - 2(0.0050) = 0.050$$

3. (a) $K_c = \dfrac{[H_2O]^2}{[H_2]^2}$

(b) $[H_2] = 0.50$; $[H_2O]^2 = K_c \times [H_2]^2 = 0.25 \times 0.25$
$$[H_2O] = 0.25$$

4. In each case, reaction proceeds to the right.

5. Let the initial concentration of HI be 1.00 M. Then $[HI] = 0.75$. The concentration of HI has decreased by 0.25 M. The concentrations of H_2 and I_2 must have increased by half as much, or 0.125 M. Hence:
$$K_c = (0.125)^2 / (0.75)^2 = 0.028$$

SELECTED READINGS

For a review of some math (e.g., quadratic equations), see the manuals listed in the Preface.

For additional help and practice at problem solving, including exercises dealing specifically with equilibrium, see the problem manuals listed in the Preface.

14

RATES OF REACTION

QUESTIONS TO GUIDE YOUR STUDY

1. Knowing what reactions should occur (those for which K_c is large), can you now say when a reaction will begin, how fast it will go, and when it will stop?

2. What properties of a system could you observe to see how fast a reaction is occurring? What kinds of measurements would you make? How can you measure the rates of very fast and very slow reactions?

3. How does the rate of a reaction depend on conditions such as temperature, pressure, and the physical state of the reactants?

4. How does the rate of a reaction depend on concentration? How do you experimentally arrive at such a rate expression?

5. How do you account for a particular rate expression in terms of molecular rearrangements? (The sum of these stepwise reactions constitutes the reaction mechanism.) Is there more than one possible mechanism?

6. How do you account for the fact that most reactions speed up as the temperature is raised? Can you quantitatively relate rate and temperature? Recall that kinetic theory relates molecular speeds and energies to temperature.

7. How must the rate of a reaction change as equilibrium is approached?

8. How can you change or control the rate of a reaction? How are reaction rates controlled or altered in the kitchen? In a living organism?

9. Can you predict or calculate the rate, or write the rate expression, for a particular reaction (as you might calculate ΔH) without ever carrying out that reaction?

10. In what ways are rate expressions useful? What, if anything, do they let you say about the feasibility of carrying out a particular reaction?

YOU WILL NEED TO KNOW

Concepts

1. How to interpret the sign of ΔH. See Chapter 6.
2. How to interpret the distribution of speeds and energies in a gas; how the distribution changes with temperature. See Chapter 7.

Math

1. How to calculate logarithms and antilogs — see Appendix 4.
2. How to work with concentrations (M). See Chapter 4.
3. How to work with the graph and equation for a straight line; how to interpret the slope of a line. See the readings listed in the Preface.

BASIC SKILLS

1. **Given the order of a reaction,**
   a. **write the rate expression.**

Example 14.1 illustrates this skill. Note that you need to distinguish between the order with respect to an individual reactant and the overall order of a reaction. See the discussion in the text preceding and directly following the example.

   b. **as well as the rate at a given concentration, calculate the rate constant k.**

This consists in simply rearranging the rate expression to solve for k. For example, for a first order reaction: $k = rate/conc$. This is shown in the second part of Example 14.1.
A related skill is to calculate the rate at a given concentration, knowing the rate expression (or order) and the value of k. See the third part of Example 14.1.

   2. **Determine the order of a reaction, given the concentration of reactant as a function of time.**

See Example 14.3. Essentially what is done here is to determine whether the data fits the rate expression for first order reactions, or the expression for some other order. Inspection of the concentration versus time data to see whether it fits the concentration-time relation, the half-life relation, or gives a linear plot – all of these are summarized by Table 14.2.

3. **Use the concentration-time relation for the first order reaction (Equation 14.8) to obtain:**
a. **the concentration of reactant after a given time, knowing the original concentration and the rate constant;**
b. **the time required for the concentration to drop to a particular value, given the original concentration and the rate constant.**

The use of the first order relation for these purposes is illustrated in Example 14.2, parts a and b. Notice that there is some advantage in retaining the quantity $Log(X_0/X)$ as a unit in your calculations. Since X_0 is always greater than X, $Log(X_0/X)$ is always a positive quantity. Most students would prefer to avoid using negative logs whenever possible.

4. **Given either the half-life or the rate constant for a first order reaction, calculate the other quantity.**

The relationship between these two quantities is derived and then applied in Example 14.2, part c. Note that for a first order reaction $t_{1/2}$ is independent of initial concentration. Note too that $t_{1/2}$ is inversely proportional to k. If a reaction is fast (large k), the half-life is short.

5. **Given two of the three quantities, E_a (activation energy for the forward reaction), E_a' (activation energy for the reverse reaction), and ΔH, calculate the third quantity.**

The appropriate equation here is 14.10. Its use is described in the body of the text and again here.

- -

For a certain reaction, $\Delta H = -80$ kJ and $E_a = 105$ kJ. What is the value of E_a', the activation energy of the reverse reaction? _____
Solving Equation 14.10 for E_a':

$$E_a' = E_a - \Delta H = 105 - (-80) = 185 \text{ kJ}$$

- -

6. **Use the "two point" equation (14.15) relating k to T to obtain**
a. **the rate constant at T_2, given its value at T_1 and the activation energy;**

  b. the activation energy, given rate constants at two different temperatures;

  c. the temperature at which k will have a specific value, given E_a, and k_1 at T_1.

Skills 6a and 6b are illustrated in Example 14.4, parts a and b. The following example illustrates Skill 6c.

  The activation energy for a certain reaction is 26.4×10^3 J. The rate constant is 1.00×10^{-2} min^{-1} at 27°C. At what temperature will k be 2.50×10^{-2} min^{-1}? _____

  Substituting into Equation 14.15:

$$\text{Log}_{10} \frac{2.50 \times 10^{-2}}{1.00 \times 10^{-2}} = \frac{26.4 \times 10^3 \, (T_2 - 300)}{(19.14)(300)(T_2)}$$

and dividing through,

$$\text{Log } 2.50 = 4.60 \times \frac{(T_2 - 300)}{T_2}$$

$$0.398 = 4.60 \times \frac{(T_2 - 300)}{T_2}$$

$$T_2 = 329 \text{ K or } 56°C$$

---

  7. **Determine whether a proposed mechanism for a reaction is consistent with the observed rate expression.**

  The general principle here is that the form of the rate expression is determined by the slowest step of the mechanism. This step often involves a very reactive intermediate which is not one of the original reactants. Such is the case with the I atom in Example 14.5. The rate expression observed in the laboratory cannot involve such intermediates. In the example quoted, the rate is expressed in terms of the concentrations of the two net reactants, H_2 and I_2.

SELF-TEST

True or False

1. The rate of reaction ordinarily decreases with time. ( )

2. If the rate of reaction doubles when the concentration is doubled, the reaction must be first order. ( )

3. In general, very fast reactions have small activation energies. ( )

4. The activation energy for a reaction can be obtained by finding the difference in enthalpy between reactants and products. ( )

5. A given molecule in a sample of gas may undergo very frequent collisions, but most of them will be ineffective. ( )

6. The order of reactant A in a mechanistic step is the same as the coefficient of A in the equation for the step. ( )

7. The reaction $H_2(g) + I_2(g) \rightarrow 2\ HI(g)$ is first order in both H_2 and I_2. Consequently, the mechanism must involve a simple collision between H_2 and I_2 molecules. ( )

Multiple Choice

8. The rate of a chemical reaction usually varies with ( )
   (a) concentration   (b) temperature
   (c) time   (d) all of these

9. For a certain decomposition the rate is 0.30 M/s when the concentration of reactant is 0.20 M. If the reaction is second order, the rate (M/s) when the concentration is 0.60 M will be ( )
   (a) 0.30   (b) 0.60
   (c) 0.90   (d) 2.7

10. In a certain first order reaction the half-life is 20 min. The rate constant k in min^{-1} is about ( )
    (a) 0.035   (b) 0.35
    (c) 13.9   (d) cannot tell

11. First order rate constants for two reactions A and B are $0.024\ h^{-1}$ and $0.13\ h^{-1}$ in that order. How do the half-lives compare? ( )
    (a) $t_{1/2}$ of A is the longer   (b) $T_{1/2}$ of A is the shorter
    (c) they are the same   (d) they cannot be
                                             compared

12. For a first order reaction, a straight line is obtained if you   ( )
plot
    (a) log conc. vs. time   (b) conc. vs. time
    (c) 1/conc. vs. time   (d) log conc. vs. 1/time

13. The activation energy of a certain reaction is 15 kJ. The   ( )
activation energy for the reverse reaction is
    (a) −15 kJ   (b) >15 kJ
    (c) <15 kJ   (d) cannot tell

14. The effectiveness of a catalyst depends upon its ability to   ( )
    (a) decrease the activation energy
    (b) increase K_c
    (c) increase reactant concentration
    (d) increase temperature

15. The principal reason for an increase in reaction rate with an   ( )
increase in temperature is
    (a) molecules collide more frequently at high temperatures
    (b) the pressure exerted by reactant molecules increases
         with T
    (c) the activation energy decreases with an increase in T
    (d) the fraction of high-energy molecules increases with T

16. For the chain reaction between H_2 and F_2 to form HF, the   ( )
step $H + F \rightarrow HF$ represents
    (a) chain initiation
    (b) chain propagation
    (c) chain termination
    (d) the overall mechanism of the reaction

17. Which of the following statements is true for all zero order   ( )
reactions?
    (a) the activation energy is very low
    (b) the concentration of reactant does not change with
         time
    (c) the rate constant, k, is zero
    (d) the rate is independent of time

18. The following mechanism is proposed for the oxidation of   ( )
iodide ion, I^-, to iodine:

$$NO + \tfrac{1}{2} O_2 \rightarrow NO_2$$

$$NO_2 + 2\,I^- + 2\,H^+ \rightarrow NO + I_2 + H_2O$$

A catalyst in this reaction is
    (a) NO   (b) I^-
    (c) O_2 (d) H^+

19. To determine the order with respect to I⁻ in the reaction ( )
2 H⁺(aq) + 3 I⁻(aq) + H₂O₂(aq) → 2 H₂O + I₃⁻(aq), you would
prepare several solutions differing in
    (a)   conc. I⁻
    (b)   conc. H⁺, conc. H₂O₂
    (c)   conc. I⁻, conc. H⁺, conc. H₂O₂
    (d)   none of the above; the order is third, as given by the
         coefficient in the balanced equation

20. When concentration of reactant is plotted against time, the ( )
rate of the forward reaction at time t is the
    (a)   intercept
    (b)   negative of the slope at t
    (c)   same as the rate constant for the forward reaction
    (d)   same as the rate of the reverse reaction

Problems

1. At 300 K the first order rate constant for the decomposition of
N_2O_5 is 4.0×10^{-4} s⁻¹.
    (a)   Write the rate expression for the reaction.
    (b)   Determine the rate of decomposition of N_2O_5 when its
         concentration is 1.8×10^{-3} M.

2. The gas phase reaction $SO_2Cl_2 \rightarrow SO_2 + Cl_2$ is found to be first
order with k = 2.20×10^{-5} s⁻¹ at 320°C.
    (a)   How long will it take for half of a 10.0 g sample of SO_2Cl_2
         to decompose at this temperature?
    (b)   Calculate the fraction of a 10.0 g sample of SO_2Cl_2 that will
         remain unreacted after two hours at 320°C.

3. For the reaction A → 2 B, ΔH = 64.0 kJ and the activation energy
of the reverse reaction is 22.0 kJ. Calculate
    (a)   the activation energy for the forward reaction.
    (b)   the rate constant for the forward reaction at 45°C if k = 1.06
         $\times 10^{-5}$ min⁻¹ at 0°C.

4. The mechanism for the decomposition of ozone is:

$$O_3(g) \rightleftharpoons O_2(g) + O(g) \quad \text{fast}$$
$$O(g) + O_3(g) \rightarrow 2\,O_2(g) \quad \text{slow}$$

Write a rate expression for this reaction which does *not* involve the
concentration of O atoms, an unstable intermediate.

5. A certain author writes about 1800 words a day under normal conditions (37°C). His editor makes a few phone calls, applying a little heat, and the author is now seen feverishly (39°C) at work, cranking out 3500 words a day. Estimate the energy barrier separating the undisciplined author from his minor masterpiece.

SELF-TEST ANSWERS

1. **T** (Rate decreases as reactant concentrations drop.)
2. **T** (Rate = k(conc.), with a direct proportion.)
3. **T** (So that a large number of reactant molecules have sufficient energy for effective collisions.)
4. **F** (ΔH is the *difference* in activation energies for forward and reverse reactions.)
5. **T** (For a collision to be effective in bringing about reaction, the molecules must have at least the energy E_a. Most molecules do not have this much energy.)
6. **T** (This is the only time you can be sure of the order with respect to A. The coefficient of A in the overall equation does not necessarily give the order.)
7. **F** (You cannot deduce the mechanism unambiguously from the rate expression. Additional evidence is needed.)
8. **d**
9. **d** (If the concentration increases by a factor of 3, the rate increases by a factor of $(3)^2 = 9$.)
10. **a** ($k = 0.693/t_{1/2}$.)
11. **a** ($t_{1/2}$ is inversely related to the rate constant.)
12. **a** (See Table 14.2.)
13. **d** (Need to know ΔH, and then to use Equation 14.10.)
14. **a** (By providing an alternate mechanism.)
15. **d** (The increase is exponential. See Chapter 7 for effect of T upon the distribution of energies.)
16. **c** (Nothing further happens to HF, the product.)
17. **d** (Except when reactants are finally gone.)
18. **a** (No net change in the NO.)
19. **a** (A general principle in any experiment; to determine the effect of factor X on a system, change only X.)
20. **b** (See the discussion at the end of Section 14.1.)

Solutions to Problems

1. (a) Rate = k (conc N_2O_5)
   $-\Delta$(conc N_2O_5)/Δt = 4.0×10^{-4} s^{-1} (conc. N_2O_5)
   (b) Rate = 4.0×10^{-4} s^{-1} (1.8×10^{-3} M)
   = 7.2×10^{-7} M/s

2. (a) $t_{1/2}$ = 0.693/k
   = $0.693/(2.20 \times 10^{-5}$ $s^{-1})$ = 3.15×10^4 s
   (b) $\text{Log}(10.0/X)$ = (2.20×10^{-5}) (7200)/2.30 = 0.0689
   $10.0/X$ = 1.172
   The fraction unreacted = $X/10.0$ = 1/1.172 = 0.853

3. (a) $\Delta H = E_a - E_a'$
   $E_a = \Delta H + E_a'$ = 64.0 + 22.0 = 86.0 kJ
   (b) $\text{Log}(k_2/k_1)$ = $\dfrac{E_a}{2.30(8.31)} \times \dfrac{T_2 - T_1}{T_2 T_1}$

   = $\dfrac{86.0 \times 10^3}{2.30(8.31)} \times \dfrac{45}{318(273)}$ = 2.33

   $k_2/(1.06 \times 10^{-5})$ = 2.15×10^2
   k_2 = 2.28×10^{-3} min^{-1}

4. Rate = $k_2 \times$ (conc. O) $\times$ (conc. O_3), where k_2 is the rate constant
   for the second reaction.

   But: $K_1 = \dfrac{(\text{conc. } O_2) \times (\text{conc. O})}{(\text{conc. } O_3)}$, where K_1 is the equilibrium
   constant for the first reaction.

   Solving for conc. O and substituting in the rate expression:

   conc. O = $K_1 \times \dfrac{(\text{conc. } O_3)}{(\text{conc. } O_2)}$; rate = $K_1 \times k_2 \times \dfrac{(\text{conc. } O_3)^2}{(\text{conc. } O_2)}$

5. $\text{Log}(3500/1800)$ = $\dfrac{E_a}{2.30(8.31)} \times \dfrac{2}{312(310)}$

   E_a = 270 kJ (Rather large — it's amazing anything gets done!)

SELECTED READINGS

An alternative discussion, at about the same level:

King, E.L., *How Chemical Reactions Occur: An Introduction to Chemical Kinetics and Reaction Mechanisms*, New York, W.A. Benjamin, 1964.

Mechanisms of reactions are discussed in:

Edwards, J.O., From Stoichiometry and Rate Laws to Mechanisms, *Journal of Chemical Education* (June 1968), pp. 381–385.

Jones, M., Jr., Carbenes, *Scientific American* (February 1976), pp. 101–113.

Wolfgang, R., Chemical Accelerators, *Scientific American* (October 1968), pp. 44–52.

Connecting reaction rates and equilibrium:

Guggenheim, E.A., More about the Laws of Reaction Rates and Equilibrium, *Journal of Chemical Education* (November 1956), pp. 544–545.

Mysels, K.J., The Laws of Reaction Rates and Equilibrium, *Journal of Chemical Education* (April 1956), pp. 178–179.

Some special effects and problems in kinetics:

Bunting, R.K., Periodicity in Chemical Systems, *Chemistry* (April 1972), pp. 18–20.

Kolb, D., Catalysis, *Journal of Chemical Education* (November 1979), pp. 743–747.

Tamaru, K., New Catalysts for Old Reactions, *American Scientist* (July-August 1972), pp. 474–479.

Winfree, A.T., Rotating Chemical Reactions, *Scientific American* (June 1974), pp. 82–95.

Math background is reviewed in Appendix 4 (logarithms) as well as the readings listed in the Preface.

15

THE ATMOSPHERE

QUESTIONS TO GUIDE YOUR STUDY

1. What are some of the bulk properties of the atmosphere? What are its dimensions and mass?

2. What is the overall composition of the atmosphere? How does it vary with weather and geographical location? What kinds of experiments supply this information?

3. How are useful compounds such as ammonia and nitric acid made from the components of the atmosphere?

4. How does the composition of the atmosphere change with height? Why does it change?

5. What reactions do the components of the atmosphere take part in, within the atmosphere itself, as well as at the interface of atmosphere and earth and between the atmosphere and living organisms?

6. What are the rates and extents of these reactions?

7. What substances can be considered as atmospheric contaminants? Where do they come from? Where do they go? What effects do they have on atmospheric properties? On life and other processes?

8. How do you test for pollutants? How do you specify their concentrations? How do you control them? How do you prevent them?

9. How do you measure the extent of pollution and its change with time? (Is pollution increasing?)

10. What are some of the reactions by which pollutants can be removed? What are their limitations? Their costs?

YOU WILL NEED TO KNOW

Concepts

1. How to interpret the sign of ΔH — Chapter 6.
2. What is meant by equilibrium vapor pressure — Chapter 11.
3. How to interpret an equation for a step in a mechanism — Chapter 14.

Math

1. How to write simple balanced equations, including those for acid-base reactions. See Chapter 4.

2. How to work with equilibrium constants, discussed in Chapter 13. (See the skills listed below.)

3. How to relate the wavelength and energy of light (see Equation 8.2 in Chapter 8); what wavelengths correspond to various regions of the spectrum. See Chapter 8 in this guide.

4. How to predict, qualitatively as well as quantitatively, the effects of changes in conditions (T, P, concentration, catalyst) on the extent of reaction (Chapter 13) and rate of reaction (Chapter 14). Particularly useful is Le Chatelier's Principle — Section 13.6.

5. How to use the rate law expressions of Chapter 14. (See the skills listed below.)

BASIC SKILLS

Most of the skills and concepts used in this chapter have already been discussed. Consider the skills reviewed in the examples of this chapter.

— *Relate the mole fractions (or mole percents) of the various species in a mixture.* To work Example 15.1 you need to realize that the sum of all the mole percents is 100%. Similarly, the sum of all mole fractions must be 1.

— *Predict the effect of a change in reaction conditions on*

   a. *the extent, or yield, of a reaction.* Part of Example 15.2 can be worked simply by applying Le Chatelier's Principle. See Section 13.6 and Skill 5 in Chapter 13 of this guide.

   b. *the rate of a reaction.* Part of Example 15.2 can be considered as a simple review of Chapter 14.

— *Quantitatively relate the rate of reaction to temperature. Example* 15.4 requires the use of Equation 14.15. See Skill 6 in Chapter 14.

— *Relate the rate expression for a reaction to the rate constant, the rate, and concentrations.* Examples 15.5 and 15.7 ask you to determine one of these, given all the others. See Skill 1b in Chapter 14.

— *Write the rate expression for a stepwise reaction in a mechanism.* See the discussion of Section 14.7 in the text as well as Skill 7 in Chapter 14 of this guide. Example 15.5 illustrates this skill.

— *Given K_c and all but one equilibrium concentration, calculate the remaining one.* Example 15.6 illustrates the application of Skill 3a in Chapter 13. Example 15.8 requires a related calculation.

Skills new to this chapter include the following.

1.   Given two of the three quantities, relative humidity, partial pressure of $H_2O(g)$ in the air, and the vapor pressure, calculate the third quantity.

Calculations of this type follow directly from the definition of relative humidity. See Example 15.3 for an illustration.

2.   Discuss some of the chemistry of the atmosphere.

This is a simple way of saying that there is a considerable amount of descriptive chemistry in this chapter. In particular, some of the things you should be able to do include:
— Write equations and describe the reaction conditions for the preparation of NH_3 (Haber process); HNO_3 (Ostwald process); and H_2SO_4 (contact process).
— Describe some of the species present in the atmosphere
   a.   at various altitudes, and
   b.   as a result of pollution.
You should be able to write equations for some reactions; describe the conditions; describe the sources and the remedies tried.

SELF-TEST

True or False

1.   The mole percent of N_2 in the atmosphere at sea level is   (   ) about 78%.

2.   Of all the gases in the atmosphere, the least reactive is   (   ) nitrogen.

3.   The inertness of N_2 is explained in terms of its very strong   (   ) double bond.

4.   When a sample of air is allowed to warm up without   (   ) changing the total water content, the relative humidity increases.

5.   As you move higher in the atmosphere, the ratio (conc. O):   (   ) (conc. O_2) is expected to increase.

6.   In order to lower the concentration of NO in automobile   (   ) exhausts, it is desirable to increase the temperature at which the fuel is burned.

7.   A catalyst is employed in the Haber process so as to increase   (   ) the extent of reaction.

Multiple Choice

8. A certain synthetic atmosphere is 80 mol % He, 19 mol %   (   )
O_2, and 1 mol % N_2. In ppm, the concentration of N_2 is
    (a) 1
                     (b) 1×10^4
    (c) 1×10^6
             (d) 1×10^8

9. In the Haber process for making ammonia, high pressures   (   )
are used to increase the
    (a) yield
                     (b) rate
    (c) yield and rate
          (d) K_c and rate

10. The major commercial use for NH_3 is the preparation of   (   )
    (a) fertilizer
             (b) explosives
    (c) plastics
               (d) steel

11. Which of the following is a starting material in the Ostwald   (   )
process?
    (a) NH_3
              (b) NO
    (c) NO_2
              (d) HNO_3

12. The conversion of NO to NO_2 by the reaction 2 NO(g) +   (   )
$O_2 (g) \rightarrow 2\ NO_2 (g)$, $\Delta H = -105$ kJ, is most extensive at
    (a) low T, low P
        (b) low T, high P
    (c) high T, high P
      (d) high T, low P

13. The major culprit in the formation of acid rain is   (   )
    (a) hydrocarbons
        (b) carbon monoxide
    (c) nitrogen oxides
      (d) sulfur oxides

14. The removal of SO_2 from stack gases makes use of the   (   )
reaction
    (a) $SO_2 (g) \rightarrow S(s) + O_2 (g)$
    (b) $SO_2 + \frac{1}{2} O_2 (g) \rightarrow SO_3 (g)$
    (c) $SO_2 (g) + H_2 O \rightarrow H_2 SO_3 (aq)$
    (d) $Ca^{2+}(aq) + 2\ OH^-(aq) + \frac{1}{2} O_2 (g) + SO_2 (g) \rightarrow$
           $CaSO_4 (s) + H_2 O$

15. Automobile exhaust emissions are not a major source of   (   )
    (a) NO
                   (b) CO
    (c) hydrocarbons
           (d) SO_2

16. CO emissions may be decreased by   (   )
    (a) burning fuels at higher T
    (b) mixing combustion reactants more thoroughly
    (c) passing combustion products over hot charcoal
    (d) all the above

17. The rate at which CO is converted to CO_2 by the reaction ( )
$CO(g) + \frac{1}{2} O_2(g) \rightarrow CO_2(g)$, $\Delta H = -283$ kJ, could be increased by
    (a) raising the temperature
    (b) using a catalyst
    (c) increasing the pressure
    (d) all of the above

18. Of the following fuels, which one produces the lowest ( )
concentration of pollutants under normal conditions?
    (a) coal   (b) wood
    (c) natural gas   (d) petroleum

19. The use of catalysts in auto exhaust systems is for the ( )
purpose of
    (a) oxidizing CO to CO_2
    (b) oxidizing hydrocarbons to CO_2
    (c) neither of these
    (d) both of these

20. How many of the following can be converted to strong ( )
acids? NO_2, SO_2, CO, CO_2
    (a) 1   (b) 2
    (c) 3   (d) 4

Problems

1. The partial pressure of water vapor in the air is 12.0 mm Hg (1.60 kPa) when the temperature is 24°C and the total pressure is 750 mm Hg (100 kPa). If the vapor pressure of water at this temperature is 22.4 mm Hg (2.99 kPa), what is the
    (a) relative humidity?
    (b) mole fraction of water in the air?

2. The reaction $O_3(g) + NO(g) \rightarrow O_2(g) + NO_2(g)$ is first order in both O_3 and NO. When the concentrations of O_3 and NO are both 3.0×10^{-8} M, the rate is 1.1×10^{-8} M/s. What is the value of the rate constant, k, for this reaction?

3. Consider the reaction $N_2(g) + O_2(g) \rightleftharpoons 2\,NO(g)$, for which $K_c = 0.10$ at 2000°C. If in an equilibrium system at this temperature, the concentrations of N_2 and O_2 are 4.2×10^{-3} M and 1.1×10^{-3} M in that order, what must be the equilibrium concentration of NO?

4. Table 15.3 in the text gives $\lambda = 79$ nm for the reaction:

$$N_2(g) \rightarrow N_2^+(g) + e^-$$

Calculate ΔE for this reaction in kilojoules per mole.

$$\Delta E(kJ/mol) = \frac{1.196 \times 10^5}{\lambda \ (nm)}$$

5. Write an equation for each step in the Ostwald process. Describe the reaction conditions.

SELF-TEST ANSWERS

1. **T** (What is the mol % O_2 in air? See Table 15.1.)
2. **F** (Argon and the other noble gases.)
3. **F** (N_2 is a very stable *triple*-bonded species. Write the Lewis structure.)
4. **F** (More vapor could be present at the higher T, since vp increases with T — as discussed in Chapter 11.)
5. **T** (Where does the energy come from that is required to break the O_2 bond?)
6. **F** (The formation of NO is endothermic and so is more extensive at higher T. Apply Le Chatelier's Principle — Section 13.6.)
7. **F** (A catalyst alters only the rate of reaction — Section 14.5.)
8. **b** (ppm = $10^6 \times$ mole fraction = $10^6 \times 0.01$.)
9. **c** (Higher P favors fewer moles of gas; it also effectively increases concentration, therefore the rate.)
10. **a** (See the discussion in Section 15.1.)
11. **a** (This is oxidized to NO, then NO to NO_2. HNO_3 is the end product.)
12. **b** (Apply Le Chatelier's Principle.)
13. **d** (See the discussion in Section 15.4.)
14. **d** (Reaction 15.27 in the text. Note that oxidation-reduction, acid-base, and precipitation are all involved here!)
15. **d** (SO_2 comes mainly from burning sulfur-containing coal and oil in power plants.)
16. **b** (How do the others favor CO formation?)
17. **d** (Chapter 14.)
18. **c**
19. **d** (See the discussion in Section 15.4.)
20. **b** (The conversions involve addition to water — to give HNO_3; oxidation to SO_3 and then addition to water — to give H_2SO_4.)

Solutions to Problems

1. (a)   R.H. = 100 $\times$ 12.0/22.4 = 53.6% (or, 100 $\times$ 1.60/2.99)
    (b)   X will be the same as P_{H_2O}/P_{tot} = 12.0/750 = 0.0160
           (or, using kPa: 1.60/100 = 0.0160)

2. $k = \dfrac{\text{rate}}{(\text{conc. } O_3)(\text{conc. NO})} = \dfrac{1.1 \times 10^{-8} \text{ M/s}}{(3.0 \times 10^{-8} \text{ M})^2} = 1.2 \times 10^7 \text{ M}^{-1} \text{ s}^{-1}$

3. $K_c = 0.10 = [NO]^2/[N_2][O_2] = [NO]^2/(4.2 \times 10^{-3})(1.1 \times 10^{-3})$
    $[NO]^2 = 4.6 \times 10^{-7}$
    $[NO] = 6.8 \times 10^{-4}$

4. $\Delta E = 1.196 \times 10^5/(79) = 1.5 \times 10^3$ kJ/mol

5. These are Reactions 15.4–15.6 in the text.
    Oxidation of NH_3, by burning it in air:
    $4 NH_3(g) + 5 O_2(g) \rightarrow 4 NO(g) + 6 H_2O(g)$, at high T, catalyst
    Mixing the product NO with air, oxidizing it further:
    $2 NO(g) + O_2(g) \rightarrow 2 NO_2(g)$, at lower T
    Passing NO_2 through water (and recycling the NO formed):
    $3 NO_2(g) + H_2O(1) \rightarrow 2 HNO_3(aq) + NO(g)$
    The aqueous solution of HNO_3 is distilled in the presence of H_2SO_4
    to obtain nitric acid.

SELECTED READINGS

Looking at the atmosphere and its relation to the rest of our planet (or some other planet), setting a perspective:

Herbig, G.H., Interstellar Smog, *American Scientist* (March-April 1974), pp. 200–207.

Huntress, W.T., Jr., The Chemistry of Planetary Atmospheres, *Journal of Chemical Education* (April 1976), pp. 204–208.

Kellogg, W.W., The Sulfur Cycle, *Science* (February 11, 1972), pp. 587–595.

Kenyon, D.H., *Biochemical Predestination*, New York, McGraw-Hill, 1969.

Leovy, C.B., The Atmosphere of Mars, *Scientific American* (July 1977), pp. 34–43.

Lewis, J.S., The Atmosphere, Clouds and Surface of Venus, *American Scientist* (September-October 1971), pp. 557–566.

O'Nions, R.K., The Chemical Evolution of the Earth's Mantle, *Scientific American* (May 1980), pp. 120–133.

The Biosphere, *Scientific American* (September 1970).

Turner, B.E., Interstellar Molecules, *Scientific American* (March 1973), pp. 51–69.

More specifically oriented toward ecological problems:

Carbon Dioxide: Atmospheric Overload, *SciQuest* (April 1980), pp. 18–22.

Carbon Monoxide: Natural Sources Dwarf Man's Output, *Science* (July 28, 1972), pp. 338–339.

Fennelly, P.F., The Origin and Influence of Airborne Particulates, *American Scientist* (January-February 1976), pp. 46–56.

Hanst, P.L., Noxious Trace Gases in the Air, *Chemistry* (January-February 1978), pp. 8–15.

Likens, G.E., Acid Rain, *Scientific American* (October 1979), pp. 43–51.

Stoker, H.S., *Environmental Chemistry: Air and Water Pollution,* Glenview, Ill., Scott, Foresman, 1972.

Some important natural and industrial processes are considered in:

Brill, W.J., Biological Nitrogen Fixation, *Scientific American* (March 1977), pp. 68–81.

Chilton, T.H., *Strong Water: Nitric Acid, Its Sources, Methods of Manufacture, and Uses,* Cambridge, Mass., MIT Press, 1968.

Cook, G.A., *Survey of Modern Industrial Chemistry,* Ann Arbor, Mich., Ann Arbor Science, 1975.

Rochow, E.G., *Modern Descriptive Chemistry,* Philadelphia, W.B. Saunders, 1977.

Safrany, D.R., Nitrogen Fixation, *Scientific American* (October 1974), pp. 64–80.

16

SOLUTES IN WATER

QUESTIONS TO GUIDE YOUR STUDY

1. What happens at the molecular level when a substance dissolves in water? What heat flow accompanies such a process and how is it explained?

2. What. properties determine the extent to which one substance dissolves in another? Are there correlations that can be made with particle structure and intermolecular forces?

3. How do the properties of a solution compare to those of the pure solvent? How do they depend on the nature and concentration of solute? Is there, as for mixtures of gases, a workable model with simple laws?

4. How do you quantitatively describe the composition of a solution? How would you experimentally determine the concentration of a solute in water solution? How would you show that a given solution is saturated? or, supersaturated? How would you prepare such solutions?

5. How do changes in conditions, such as T and P, affect solubility? Can these effects be predicted?

6. How do you write chemical equations for the formation of a solution? of a precipitate? How do you interpret these equations in stoichiometric calculations?

7. What can you say about the rates of reactions in solution, or about their extents? What is the particle nature of a precipitate?

8. What are some of the practical applications of solubility principles in and out of the laboratory?

9. What effect does a solute have on the vapor pressure of water? the boiling point? the freezing point?

10. How do water solutions of ionic solutes differ in their properties from water solutions containing molecular solutes?

YOU WILL NEED TO KNOW

Concepts

1. Some of the more common ions — names, formulas, and charges. See Table 3.2 and Figure 3.3 in the text.

2. How to recognize substances as being molecular (polar and nonpolar) or ionic; how to predict the kinds and relative magnitudes of attractive forces between molecules and between ions. See Skills 2 and 3 in Chapter 11 of this guide.

Math

1. How to write balanced equations for simple chemical reactions. See Sections 4.1 and 4.5 (particularly the latter) for a discussion of reactions involving species in water solution.
2. How to use balanced equations in working stoichiometric problems. See Sections 4.2 and 4.3 as well as 4.5 again.
3. How to work with the concentration unit M (mol/ℓ or mol/dm³) — See Section 4.4 in the text.
4. How to apply Le Chatelier's Principle in predicting the effect of a change in conditions on the extent of a reaction. See Skill 5 in Chapter 13 of this guide.

BASIC SKILLS

As you probably noticed above, this chapter relies heavily on material discussed in Chapter 4. In fact, precipitation reactions were introduced there and now serve as the core of this chapter. Titration also was introduced in that earlier chapter. It might be very useful to carefully work through some of the examples and skills of Chapter 4 at this point.

1. **Given the formula of a substance, decide whether it is likely to be an electrolyte or a nonelectrolyte in water solution.**

Basic to this skill is the ability to recognize the solute as a molecular substance, polar or nonpolar, or as an ionic solid. The ionic solid will always give ions in solution, provided simply that the solid is indeed soluble. If the solute is nonpolar, chances are it will dissolve very little and give only molecular, nonconducting solute particles. A few of the polar molecular solutes, such as HCl and HBr, dissolve to give ions in solution. Table 16.2 gives several representative examples.

You should then be able to compare water solutions of any two solutes, at the same concentration, as to electrical conductivity and as to their freezing and boiling points, vapor pressures, and osmotic pressures. See the skills below.

2. **Predict the relative solubilities of different solutes in water.**

The principle here is that solubility is enhanced by similarity in the type and strength of intermolecular forces. (This means that you must first recognize the nature of the particle structure and the forces between particles.) Other than this general principle, you will need to **know the solubility rules** (Table 16.3) in order to decide on the solubility of a given ionic solid. See the catalog of problems for illustration.

3. **Predict the effect on solubility of a change in T or P.**

The effect of a change in pressure can be ignored except in the case of a gaseous solute. There, higher partial pressure of the solute above the solution increases solubility. The effect of temperature changes can be predicted from the sign of ΔH for the solution process. If ΔH is positive, solubility increases with temperature; if ΔH is negative, solubility decreases with increasing temperature. Note that ΔH is usually, but not always, positive for a solid solute; it is negative for a gaseous solute.

4. **Write a balanced (net ionic) equation for the formation of a**
   a. **solution;**
   b. **precipitate.**

Unless you are given the formulas for all species as they occur in the reactants and products, you will first need to apply Skill 1 above. Then, knowing these formulas, you need to consider two other points: include only those species that take part in the overall, net reaction. (Exclude those that start in solution and remain there.) Then, balance the equation: include all states, using the symbols (aq) and (s).

Example 16.1 illustrates the formation of solutions of ionic electrolytes. Example 16.3 applies part b of this skill.

Note the discussion in Example 16.3. You must consider the two possible combinations of cation and anion. You must then know the solubility rules in order to decide whether either combination is insoluble. Example 16.2 illustrates the application of the solubility rules to specific combinations of anion and cation (simple ionic compounds).

5. **Given or having derived the equation for a precipitation reaction, relate the amounts of reactants and products.**

This skill is not new to this chapter; you saw it in Chapter 4. Example 16.4 is thus a review of this material, and is not unlike Example 4.9 in the principles being applied. The exercise that follows Example 16.4 is also like that following 4.9. Both exercises describe the basic nature of titration.

If you need to review the concepts and calculations involving concentration, see Section 4.4 in the text.

6.   Use the limting laws pertaining to boiling point elevation (Equation 16.13) and freezing point depression (Equation 16.14) for solutions of nonelectrolytes to relate:
a.   the boiling point (freezing point), molality, and k_b (k_f).
b.   the boiling point (freezing point), molality, mode of ionization (or value of i), and k_b (k_f).

Example 16.5 illustrates a; you are asked to determine the freezing point of the solution, given k_f and masses of solute and solvent. In order to work this example, it is necessary to determine the molality of solute. This follows directly from the definition: m = mole solute/kilogram solvent.

Example 16.6 applies part b of this skill. There, you are asked to determine the molecular mass of solute, given all the necessary data to apply Equation 16.14 in a slightly different form. In particular, rewriting molality: m = (g solute/GMM solute)/kg solvent, where GMM is the molecular mass, in grams. So Equation 16.14 becomes:

$$\Delta T_f = k_f \times \frac{\text{g solute/GMM solute}}{\text{kg solvent}}$$

The exercise following Example 16.6 is a similar problem. See the catalog of problems for additional practice.

7.   Use the limiting laws (Skill 6) as modified for solutions of electrolytes to relate:

a.   the boiling point (freezing point), molality, and k_b (k_f).
b.   the boiling point (freezing point), molality, mode of ionization (or value of i), and k_b (k_f).

The change in boiling point or freezing point depends primarily on the concentration of solute particles, not on their type.

-- --

Compare the freezing point lowering of 0.10 m solutions of KNO_3, $Mg(NO_3)_2$, and $Al(NO_3)_3$ to that of 0.10 m NaCl.

First, recognizing the solute particles as the ions K^+, NO_3^-, Mg^{2+}, and Al^{3+}, as well as Na^+ and Cl^-, we can make the following tabulation for one mole of solute.

	No. moles cations	No. moles anions	Total no. moles
NaCl	1	1	2
KNO_3	1	1	2
$Mg(NO_3)_2$	1	2	3
$Al(NO_3)_3$	1	3	4

We would then expect that, at the same initial concentration, the freezing point of NaCl and KNO_3 solutions should be about the same. $Mg(NO_3)_2$ should have a freezing point lowering about $3/2$ that of NaCl; $Al(NO_3)_3$ about twice that of NaCl.

_ _

It should be emphasized that such comparisons as made above apply strictly only in dilute solution, where interionic attractive forces are of minor importance. In practice, they are fairly reliable up to about 0.10 m.

Example 16.7 illustrates the second part of this skill. Another example of this skill follows.

_ _

What is the freezing point of a 1.00 m NaCl solution if i at this concentration is 1.80? _____

Substituting in Equation 16.15:

$$\Delta T_f = 1.86°C \times 1.00 \times 1.80$$
$$= 3.35°C$$

So the freezing point must be $-3.35°C$. Note that if NaCl behaved ideally at this concentration (i = 2), ΔT_f would be $2 \times 1.86 = 3.72°C$. The observed effect is *smaller* than predicted on the basis of ideal behavior (complete independence of the ions).

_ _

SELF-TEST

True or False

1. If water saturated with nitrogen at 1 atm is exposed to air, ( )
N_2 will come out of solution.

2. A supersaturated solution of N_2 in water could be prepared ( )
by bubbling N_2 through water at a higher T and then cooling.

3. A 1 m sugar solution has a total concentration of solute ( )
particles of about 1 mol/kg solvent.

4. A 1 m KCl solution should freeze at about $-2°C$.   ( )

5. The vapor pressures of two solutions at the same concen- ( )
tration, one a nonelectrolyte, the other an electrolyte, should be
about the same.

6. Any nonelectrolyte, at a concentration of 1 m, should raise ( )
the boiling point of water by about $0.52°C$.

7. The salt MNO_3 might be prepared by a precipitation ( )
reaction involving $AgNO_3$ and a solution of MCl.

Multiple Choice

8. Vapor pressure lowering is related to and can be used to ( )
help explain
    (a) boiling point elevation   (b) freezing point lowering
    (c) osmotic pressure   (d) all of these

9. When 1 mol $Mg(NO_3)_2$ is dissolved in 1 ℓ (1 dm^3) of ( )
solution, the concentration of NO_3^- is
    (a) 0.5 M   (b) 1 M
    (c) 2 M   (d) 3 M

10. Which of the following has the least effect on the solubility ( )
of a solid in water?
    (a) nature of solute   (b) temperature
    (c) pressure   (d) all are important

11. Which one of the following liquids would you expect to be ( )
the most soluble in water?
    (a) CCl_4 (b) C_8H_{18}
    (c) $CH_3-CH_2-O-CH_3$ (d) CH_3-CH_2-OH

12. Separate solutions are made of each of the following ( )
substances by dissolving 0.1 mol in 1 kg of water. Which one will
have the lowest freezing point?
    (a) CH_3OH (b) HCl
    (c) KCl   (d) K_2CO_3

13. At low concentrations, i for $Fe(NO_3)_3$ approaches the value ( )
    (a) 1   (b) 2
    (c) 3   (d) 4

14. Which kind of species should *not* be present at appreciable ( )
concentration in 0.1 M NaCl?
    (a) H_2O (b) NaCl
    (c) Na^+ (d) Cl^-

15. All chlorides are   ( )
    (a) soluble
    (b) insoluble
    (c) soluble except those of Ag^+, Hg_2^{2+}, and Pb^{2+}
    (d) insoluble except those of Ag^+, Hg_2^{2+}, and Pb^{2+}

16. When $200\ cm^3$ of 0.10 M $BaCl_2$ is added to $100\ cm^3$ of ( )
0.30 M Na_2SO_4, the number of moles of $BaSO_4$ precipitated is
    (a) 0.010   (b) 0.020
    (c) 0.030   (d) 0.20

17. When solutions of $Pb(NO_3)_2$ and Na_2SO_4 are mixed, the ( )
precipitate that forms is
    (a) $PbSO_4$ (b) $NaNO_3$
    (c) $PbSO_4$ and $NaNO_3$ (d) none

18. The net ionic equation for the reaction that occurs when ( )
water solutions of $CaCl_2$ and Na_2CO_3 are mixed is
    (a) $CaCl_2(aq) + Na_2CO_3(aq) \rightarrow CaCO_3(s) + 2\ NaCl(aq)$
    (b) $Ca^{2+}(aq) + CO_3{}^{2-}(aq) \rightarrow CaCO_3(s)$
    (c) $CaCO_3(s) \rightarrow Ca^{2+}(aq) + CO_3{}^{2-}(aq)$
    (d) $Na^+(aq) + Cl^-(aq) \rightarrow NaCl(s)$

19. The dissolving of the electrolyte HBr(1) in water is best ( )
represented by the equation
    (a) $HBr(l) \rightarrow HBr(aq)$
    (b) $HBr(g) \rightarrow H^+(aq) + Br^-(aq)$
    (c) $H^+(aq) + Br^-(aq) \rightarrow HBr(s)$
    (d) something else

20. How many of the following solids are insoluble in water? ( )
$(NH_4)_2SO_4$, $CuCl_2$, $MgCO_3$, KOH, CaS
    (a) 0   (b) 1
    (c) 2   (d) some other number

Problems

1. Predict which member of each pair is expected to be the more
soluble in water. State your reasoning.
    (a) CH_4 or CH_3OH (b) NH_4NO_3 or AgCl

2. Write equations (balanced and net ionic, of course) for any reaction occurring on mixing
    (a) dilute hydrochloric acid and lead nitrate solutions
    (b) 0.1 M solutions of NaOH and $MgCl_2$
    (c) 0.1 M solutions of NaCl and KNO_3

3. A water solution contains 12.0 g of sugar (molecular mass = 342) in 200 g of water. What is:
    (a) the molality of sugar?
    (b) the freezing point of the solution ($k_f = 1.86°C$)?
    (c) the normal boiling point of the solution ($k_b = 0.52°C$)?

4. A solid sample contains $MgSO_4$ (formula mass = 120) and NaCl (formula mass = 58.4). Treatment of an aqueous solution of 0.750 g of this sample with $BaCl_2$ solution yields 0.930 g of $BaSO_4$ (formula mass 233). Calculate:
    (a) the number of moles of $BaSO_4$ precipitated.
    (b) the percentage by mass of $MgSO_4$ in the original sample.

5. A student measures the boiling point of CS_2 to be 46.1°C. A 1.00 m solution of a certain solute in CS_2 boils at 48.5°C. When 1.5 g of solid sulfur is dissolved in 12.5 g CS_2, the boiling point is found to be 47.2°C. Determine the molecular formula of sulfur. (Atomic mass sulfur = 32.06.)

SELF-TEST ANSWERS

1. **T** (The partial pressure of N_2 in air is less than 1 atm. In fact, it is just 0.78 atm. As noted in Section 16.3, the solubility of a gas is directly proportional to its partial pressure.)

2. **F** (At lower T, and then carefully warming to the desired T. The solubility of a gas in water decreases as T increases.)

3. **T** (Sugar, a nonelectrolyte, dissolves to give 1 mol particles in solution per mole of solid.)

4. **F** ($\Delta T_f = k_f \times m \times i = 1.86(1)(2) = 3.72°C$.)

5. **F** (As for other colligative properties, the effect is larger for the electrolyte since it gives more particles in solution per mole of solute dissolving.)

6. **T** (Provided it is nonvolatile. Otherwise, the boiling point is actually lowered. See the footnote on antifreezes in Section 16.5.)

7. **T** (Mixing solutions will precipitate AgCl. Filter, and evaporate the solvent to collect MCl.)

8. **d** (See the text discussion at the beginning of Section 16.5.)
9. **c** (One mole $Mg(NO_3)_2$ gives 2 mol NO_3^-.)
10. **c** (Important only for gaseous solutes.)
11. **d** (Hydrogen bonding is present in only this solute.)
12. **d** (i = 3 for K_2CO_3; i = 2 for HCl and KCl; i = 1 for molecular CH_3OH.)
13. **d** (What kind and number of ions form per mole of solid dissolving?)
14. **b** (NaCl implies a molecule.)
15. **c** (Solubility rules, Table 16.3.)
16. **b** (Moles Ba^{2+} = 0.020; moles SO_4^{2-} = 0.030. They react in a 1:1 mole ratio. So there is an excess of SO_4^{2-}, and 0.020 mol of $BaSO_4$ precipitates.)
17. **a** (Again, you need to know the solubility rules to answer this.)
18. **b** (What would C represent?)
19. **b** (Analogous to HCl. Ions must be present in the water solution of an electrolyte.)
20. **b** (Only $MgCO_3$; the solubility rules strike again!)

Solutions to Problems

1. (a) CH_3OH (Hydrogen bonding.)
    (b) NH_4NO_3 (NH_4^+ as well as NO_3^- salts are soluble. Also, AgCl is insoluble.)
2. (a) $2\ Cl^-(aq) + Pb^{2+}(aq) \rightarrow PbCl_2\,(s)$
    (b) $2\ OH^-(aq) + Mg^{2+}(aq) \rightarrow Mg(OH)_2\,(s)$
    (c) no reaction
3. (a) $m = \dfrac{12.0\ g/(342\ g/mol)}{0.200\ kg} = 0.175$
    (b) $\Delta T_f = k_f \times m = 1.86°C(0.175) = 0.326°C$
         $T_f = -0.326°C$
    (c) $\Delta T_b = k_b \times m = 0.52°C(0.175) = 0.091°C;\ T_b = 100.091°C$
4. (a) moles $BaSO_4$ = 0.930 g × 1 mol/233 g = 3.99×10^{-3}
    (b) moles $MgSO_4$ = moles $BaSO_4$
         g $MgSO_4$ = 3.99×10^{-3} mol × 120 g/mol = 0.479 g
         % $MgSO_4$ = (0.479/0.750) × 100 = 63.9
5. $k_b = \Delta T_b/m = (48.5 - 46.1)/1.00 = 2.4°C$
    $47.2 - 46.1 = 2.4 \times \dfrac{1.5\ g/GMM}{0.0125}$
    GMM = 260 g/mol
    S_8

SELECTED READINGS

Colligative properties are correlated with vapor pressure; how is vp lowering explained? Consider:

Mysels, K.J., The Mechanism of Vapor-Pressure Lowering, *Journal of Chemical Education* (April 1955), p. 179.

Problem solving (concentrations, solution stoichiometry . . .) is illustrated in the manuals listed in the Preface.

Unusual solutions (from electrons to macromolecules) are described in:

Dye, J.L., The Solvated Electron, *Scientific American* (February 1967), pp. 76–83.
Feeney, R.E., A Biological Antifreeze, *American Scientist* (November-December 1974), pp. 712–719.

Water systems, natural and polluted, are described in:

MacIntyre, F., The Top Millimeter of the Ocean, *Scientific American* (May 1974), pp. 62–77.
Stoker, H.S., *Environmental Chemistry: Air and Water Pollution,* Glenview, Ill., Scott, Foresman, 1972.

17

ACIDS AND BASES

QUESTIONS TO GUIDE YOUR STUDY

1. What properties are characteristic of an acid; a base? How are these properties explained?

2. What happens when a substance dissolves in water to give a solution that is acidic or basic? What species are present in such solutions? Can you predict which substances will be acidic, which basic, and which neutral?

3. What is meant by the strength of an acid or base? What are the common strong acids and bases?

4. What equations do you write to describe the formation of an acidic or basic solution? What can you say about their extents and rates?

5. What role does water play in solutions that are acidic or basic? What kinds of substances can act as acids or bases outside water solutions?

6. What reactions occur between acids and bases? What, for example, constitutes a neutralization?

7. How would you follow an acid-base reaction experimentally? What would you measure?

8. How does an acid-base indicator work? How would you select one?

9. What is pH? How is it useful?

10. What are some of the applications of acid-base reactions?

YOU WILL NEED TO KNOW

Concepts

1. How to write Lewis structures for simple molecules and ions. See Skill 2 in Chapter 10 of this guide.

Math

1. How to write and balance net ionic equations. How to use them in stoichiometric calculations. Chapters 4 (Section 4.5 in the text) and 16 (see Skill 4 in Chapter 16 of this guide).

2. How to work with concentration (mol/ℓ, mol/dm^3). See Section 4.4 in the text.

3. How to determine logarithms and antilogs — see Appendix 4. Also see the discussion in Example 17.2.

4. How to write the expression for an equilibrium constant, K_c. How to predict the effect on the position of equilibrium brought about by a change in conditions, particularly concentration. See Skills 1 and 5 in Chapter 13 of the guide.

BASIC SKILLS

1. Given one of the concentrations [H$^+$] or [OH$^-$] for water or a water solution, calculate the other concentration.

These are equilibrium concentrations, expressed in moles per litre (or, mol/dm^3), and so are related by an equilibrium constant. At 25°C in water or any water solution: $K_w = [H^+] [OH^-] = 1.0 \times 10^{-14}$.

Note the inverse relation between the two concentrations. If one is somehow increased, the other must decrease. Their product remains unchanged.

Example 17.1 illustrates the use of this skill. It is used also in most of the problems and remaining worked examples in the chapter.

2. Given either [H$^+$] or [OH$^-$], calculate the pH.

The defining relationship is pH = $- \text{Log}_{10} [H^+]$. Its use is seen in Example 17.2. You should also be able to work the calculation in the reverse direction (Example 17.2b).

3. Write an equation for the dissociation of a strong acid or strong base in water solution.

You should know the few common strong acids and bases. The strong acids are: HCl, HBr, HI, HNO$_3$, H$_2$SO$_4$, and HClO$_4$. The strong bases are: the hydroxides of the metals of Groups 1 and 2 (examples: NaOH and Ba(OH)$_2$).

See Equations 17.4 and 17.6–17.7 for representative strong acids; Equations 17.13–17.14 for strong bases. Also see Equation 17.29 as an alternative expression (see the last skill below) for the formation of a solution of a strong acid.

This skill is essential to working several kinds of problems. You need to realize what species predominate, that is, what species may serve as reactants in solutions of strong acids and bases. If, for example, a solution of the strong acid HCl takes part in a reaction, it is the species $H^+(aq)$ that acts as the acid, not the molecular species HCl(aq).

4. Write an equation for the reversible reaction for:
    a.   the dissociation of a weak acid in water solution;
    b.   the reaction of a weak base with water.

Given the formula of a species that acts as a weak acid, you then know the formula of a product. It is that species *minus* one proton. (Example: the weak acid HB must form the species B^-.) Likewise, if you are given the formula of a weak base, you automatically know the formula of a product. It is that same species *plus* a proton. (Example: the weak base B^- will be converted to HB.)

Again, if you know a solution is acidic, one product *must* be H^+. If basic, the products *must* include OH^-.

Examples 17.3 and 17.4 and the exercises following them illustrate these points in detail for several acidic and basic species.

5. Predict whether a given ionic compound will give an acidic, basic, or neutral solution. Write an equation to explain your prediction.

Predictions of this sort can be made with the aid of Table 17.6 in the text. This table may seem formidable at first glance, but the principle behind it is really quite simple. The neutral anions are those derived from strong acids; the neutral cations are those derived from strong bases. Anions derived from weak acids are basic; cations derived from weak bases are acidic. There are essentially infinite numbers of ions in these two categories; those listed in the table as "basic anions" and "acidic cations" are only typical examples. Finally, there are only two oddball anions (HSO_4^- and $H_2PO_4^-$) you are likely to encounter which are acidic. They contain an ionizable proton.

As Example 17.5 implies, anions and cations can be combined in four different ways:
— neutral cation, neutral anion: neutral solution
— neutral cation, basic anion: basic solution
— acidic cation, neutral anion: acidic solution
— acidic cation, basic anion: may be acidic, basic, or neutral

The fourth type of compound cannot be classified without additional information about the relative strengths of the acidic and basic ions.

Once you have decided on the nature of the solution, whether acidic, basic, or neutral, there is the problem of writing the equation. That requires the application of Skill 4 above. If you decide it is neutral, there is no reaction to explain and no equation to write. If the solution is acidic, one product must be H^+. The other product is simply the conjugate base of the reactant.

For example, $NaHSO_4$ is acidic. The equation you would write would be:

$$HSO_4^-(aq) \rightleftharpoons H^+(aq) + SO_4^{2-}(aq)$$

In order to make a solution basic, a species must ordinarily pick up a proton from a water molecule, leaving an OH^- ion behind. Thus, to explain the fact that solutions containing NH_3, F^-, and CO_3^{2-} are basic, you would write:

$$NH_3(aq) + H_2O \rightleftharpoons NH_4^+(aq) + OH^-(aq)$$

$$F^-(aq) + H_2O \rightleftharpoons HF(aq) + OH^-(aq)$$

$$CO_3^{2-}(aq) + H_2O \rightleftharpoons HCO_3^-(aq) + OH^-(aq)$$

Skills 4 and 5, shown here together, are central to most of the work of Chapter 17 and much of Chapter 18 as well. See the catalog of problems in the text for additional practice in their use.

6. **Write an equation for the reaction of an acid with a base. Describe the solution that results as being acidic, basic, or neutral.**

Example 17.6 and the exercise that follows it apply this skill.

The nature of the equation you write depends on whether the acid and base are strong or weak.

a. Strong acid — strong base. The equation to expect is that called neutralization: $H^+(aq) + OH^-(aq) \rightarrow H_2O$. The solution would be neutral.

b. Weak acid — strong base. Here you need to realize that the predominant species in the solution of a weak acid is the undissociated species itself and not $H^+(aq)$. The other reactant is OH^-; the products are H_2O and the conjugate base of the weak acid. Typical examples include:

$$HC_2H_3O_2(aq) + OH^-(aq) \rightarrow H_2O + C_2H_3O_2^-(aq)$$

$$HCO_3^-(aq) + OH^-(aq) \rightarrow H_2O + CO_3^{2-}(aq)$$

$$NH_4^+(aq) + OH^-(aq) \rightarrow H_2O + NH_3(aq)$$

These are the equations that you would write to describe the reactions of a solution of $NaOH$ with solutions of $HC_2H_3O_2$, $NaHCO_3$, and NH_4Cl, respectively. The solutions found in each case are basic. Each contains a weak base ($C_2H_3O_2^-$, HCO_3^-, NH_3).

c. Strong acid — weak base. Here the predominant species in the solution of the weak base is the undissociated species itself and not $OH^-(aq)$.

The other reactant is H^+; the product is the conjugate acid of the weak base. For the reaction of a solution of any strong acid with solutions containing the NH_3 molecule, the S^{2-} ion or the $CO_3{}^{2-}$ ion, you would write:

$$NH_3(aq) + H^+(aq) \rightarrow NH_4{}^+(aq)$$

$$S^{2-}(aq) + 2H^+(aq) \rightarrow H_2S(aq)$$

$$CO_3{}^{2-}(aq) + 2H^+(aq) \rightarrow H_2CO_3(aq) \text{ or } CO_2(aq) + H_2O$$

The solutions resulting from these reactions contain weak acids ($NH_4{}^+$, H_2S, H_2CO_3). They are all acidic.

7. **Use titration data for an acid-base reaction to determine:**
    a.   **the concentration of an acid or base in water solution.**
    b.   **the molecular mass of an acid or a base.**

The nature of a titration was first discussed in Chapter 4, Section 4.5. See Example 4.10 as well as Example 17.7 in the current chapter of the text. These examples illustrate the calculations of part a of this skill. Except that they use concentrations, these calculations are not different from most of the other stoichiometric calculations you have done.

The second part of the skill is illustrated by Example 17.8. Note that to do this kind of calculation, you must know the equation for the reaction.

8.   **Select an acid-base indicator appropriate to a given acid-base titration.**

Your choice depends on the combination of strong or weak acid and base. (See Skill 6 above). The way the choice is made is discussed in the body of the text and illustrated in Figures 17.4 and 17.5. The range of pH over which the indicator changes color must include the pH you expect for the product solution. For example, if NaOH is tirated against NH_4Cl (strong base + weak acid), the resultant solution contains a weak base (NH_3). Choose an indicator that changes color at a pH above 7. See the problems for further illustration.

9.   **Classify a given species in a reaction as an acid or base, according to the models of Arrhenius; Brönsted and Lowry; and Lewis. Indicate the conjugate acid-base pairs.**

This skill requires only that you apply the definitions given in the text. Example 17.9 does this for Brönsted-Lowry acids and bases. It is this model in particular that is the most useful for water solutions.

To identify a species as a Lewis acid or base, it is helpful to start by writing the Lewis structure. Consider water, for example. You see from its structure

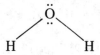

that it could donate a pair of unshared electrons. It could not, however, accept a pair of electrons, because there is no place to put them.

TABLE 1
TYPES OF REACTIONS INVOLVING ACIDS AND BASES

1. *Formation of an Acidic Water Solution*

   a. Strong acid: $HX(aq) \rightarrow H^+(aq) + X^-(aq)$
      $(X^- = Cl^-, Br^-, I^-, NO_3^-, ClO_4^-, HSO_4^-)$

   b. Weak acid: $HA(aq) \rightleftharpoons H^+(aq) + A^-(aq)$
      (A^- is any anion other than those listed in 1a.)

   Other species that act as weak acids include NH_4^+ and HSO_4^-:

   $$NH_4^+(aq) \rightleftharpoons H^+(aq) + NH_3(aq)$$
   $$HSO_4^-(aq) \rightleftharpoons H^+(aq) + SO_4^{2-}(aq)$$

   and transition metal cations, of which Zn^{2+} is typical:

   $$Zn(H_2O)_4{}^{2+}(aq) \rightleftharpoons H^+(aq) + Zn(H_2O)_3OH^+(aq)$$

2. *Formation of a Basic Water Solution*

   a. Strong base: $MOH(s) \rightarrow M^+(aq) + OH^-(aq)$; Group 1 hydroxides
      $M(OH)_2(s) \rightarrow M^{2+}(aq) + 2\,OH^-(aq)$; Group 2 hydroxides

   b. Weak base: $A^-(aq) + H_2O \rightleftharpoons HA(aq) + OH^-(aq)$
      (A^- is the anion derived from a weak acid HA.)

   The only other weak base considered here is NH_3:

   $$NH_3(aq) + H_2O \rightleftharpoons NH_4^+(aq) + OH^-(aq)$$

3. *Acid-Base Reactions*

   a. Strong acid – strong base: $H^+(aq) + OH^-(aq) \rightarrow H_2O$

   b. Strong acid – weak base: $H^+(aq) + A^-(aq) \rightarrow HA(aq)$
      $H^+(aq) + NH_3(aq) \rightarrow NH_4^+(aq)$

   c. Weak acid – strong base: $HA(aq) + OH^-(aq) \rightarrow A^-(aq) + H_2O$

Note that the reactions referred to in 1a, 2a, and 3 have very large equilibrium constants and go essentially to completion. The reactions in 1b and 2b have small equilibrium constants and occur to only a small extent. (See Chapter 18.)

As you can see, there is an unusually large number of "basic skills" in this chapter. Students often have difficulty in applying these skills. In particular, it often seems that a bewildering variety of different reactions are involved. Actually, only three major types of reactions are involved. These are summarized with the corresponding equations in Table 1. You may find that the information given in this table helps to tie together many of the skills in this chapter.

SELF-TEST

True or False

1. In a basic water solution at 25°C, $[H^+] > 10^{-7}$.   ( )

2. It is impossible to have a solution with a negative pH.   ( )

3. A solution which gives off bubbles when Na_2CO_3 is added   ( )
is likely to be acidic.

4. There are more weak acids than strong acids.   ( )

5. Solutions containing sodium acetate are expected to be   ( )
basic.

6. When equal volumes of $NH_3(aq)$ and $HCl(aq)$, both at the   ( )
same concentration, are mixed, the result is a basic solution.

7. The equivalence point in any acid-base titration is at pH 7.   ( )

Multiple Choice

8. The concentration of H^+ in a solution is 2×10^{-4} M. The   ( )
OH^- concentration (M) must be
    (a) 2×10^{-4} (b) 1×10^{-10}
    (c) 2×10^{-10} (d) 5×10^{-11}

9. The pH of the solution in Question 8 is   ( )
    (a) 3.0   (b) 3.7
    (c) 4.0   (d) 10.3

10. The solution referred to in Question 8 is: ( )
    (a) acidic   (b) basic
    (c) neutral   (d) at the equivalence point

11. A solution that is 0.10 M in HCl would have a pH of   ( )
    (a) 0.00   (b) 1.00
    (c) 7.00   (d) 13.00

12. The pH of a solution that is 0.10 M in a weak acid would be   ( )
    (a) less than 1   (b) 1.00
    (c) greater than 1   (d) impossible to say

13. Which one of the following is *not* a strong acid?   ( )
    (a) HCl   (b) HF
    (c) HNO_3 (d) $HClO_4$

14. Which one of the following is a strong base?   ( )
    (a) $Al(OH)_3$ (b) NH_3
    (c) C_2H_5OH (d) NaOH

15. Which one of the following ions would give a basic solution   ( )
upon addition to water?
    (a) NH_4^+ (b) Na^+
    (c) $C_2H_3O_2^-$ (d) NO_3^-

16. Which one of the following equations would you write to   ( )
explain the fact that a water solution of NaCl is neutral?
    (a) $Na(H_2O)_4^+(aq) \rightarrow Na(H_2O)_3(OH)(aq) + H^+(aq)$
    (b) $Na(s) + H_2O \rightarrow Na^+(aq) + OH^-(aq) + \frac{1}{2} H_2(g)$
    (c) $Cl^-(aq) + H_2O \rightarrow HCl(aq) + OH^-(aq)$
    (d) none of the above

17. The equation for the reaction of a water solution of the   ( )
weak acid HF with a solution of NaOH is best written as
    (a) $H^+(aq) + F^-(aq) \rightarrow HF(aq)$
    (b) $H^+(aq) + OH^-(aq) \rightarrow H_2O$
    (c) $HF(aq) + OH^-(aq) \rightarrow H_2O + F^-(aq)$
    (d) $HF(aq) + NaOH(aq) \rightarrow H_2O + NaF(aq)$

18. The equation for the reaction of a water solution of   ( )
ammonia with a water solution of HCl is best written as
    (a) $NH_3(aq) + H_2O \rightarrow NH_4^+(aq) + OH^-(aq)$
    (b) $NH_4^+(aq) \rightarrow NH_3(aq) + H^+(aq)$
    (c) $NH_3(aq) + H^+(aq) \rightarrow NH_4^+(aq)$
    (d) $NH_3(aq) + HCl(aq) \rightarrow NH_4^+(aq) + Cl^-(aq)$

19. In the reversible reaction $HCO_3^-(aq) + OH^-(aq) \rightleftharpoons$ ( )
$CO_3^{2-}(aq) + H_2O$, the Brönsted acids are
    (a) HCO_3^- and CO_3^{2-} (b) HCO_3^- and H_2O
    (c) OH^- and H_2O (d) OH^- and CO_3^{2-}

20. In the reaction $BF_3 + NH_3 \rightarrow F_3B:NH_3$, BF_3 accepts an ( ) electron pair and acts as
    (a) an Arrhenius base   (b) a Brönsted acid
    (c) a Lewis acid   (d) a Lewis base

Problems

1. Lemon juice has a pH of about 2.3 What is:
    (a) $[H^+]$?   (b) $[OH^-]$?

2. Write an equation to account for the fact that:
    (a) $HI(g)$ dissolves in water to give an acidic solution
    (b) a water solution of $HClO$ has a pH less than 7
    (c) $[OH^-] > 10^{-7}$ in an aqueous solution of sodium fluoride
    (d) a gas is evolved when $HCl(aq)$ is added to washing soda, Na_2CO_3.

3. Consider the reaction that occurs on mixing water solutions of HNO_2 and KOH.
    (a) Write the equation for the reaction. Describe the product solution as acidic, basic, or neutral.
    (b) Calculate the volume of 0.100 M KOH required to react with 25.0 cm^3 of 0.156 M HNO_2.

4. Which of the following species

$$NH_3, NH_4^+, SO_4^{2-}, HSO_4^-$$

can act as a
    (a) Brönsted acid?   (b) Brönsted base?

5. At $25°C$, $K_w = 1.0 \times 10^{-14}$. At $60°C$, K_w is about 10^{-13}.
    (a) What is the sign of ΔH for the ionization of water?
    (b) A certain water solution has a pH of 6.8 at $60°C$. Is it acidic, basic, or neutral?

SELF-TEST ANSWERS

1. **F** ($[H^+]$ would be *less* than that in pure water or a neutral solution.)
2. **F** (What would you calculate for the pH of 10 M HCl?)

3. **T** (A property characteristic of acids. See Problem 2d below.)
4. **T** (And more numerous weak bases than strong bases.)
5. **T** (Na^+ has no effect. $C_2H_3O_2^-$ is the anion of a weak acid; it is a weak base, tending to acquire a proton and form $HC_2H_3O_2$.)
6. **F** (The final solution contains NH_4Cl. NH_4^+ is a weak acid.)
7. **F** (This would be the case only for strong acid–strong base.)
8. **d** ($[OH^-] = K_W/[H^+] = 10^{-14}/(2 \times 10^{-4}) = 5 \times 10^{-11}$.)
9. **b** ($pH = -Log(2 \times 10^{-4}) = -(-3.7) = 3.7$.)
10. **a** (pH less than 7.)
11. **b** ($-Log\ 10^{-1}$.)
12. **c** (There would be less H^+ than in a strong acid solution of the same initial concentration, as in Question 11.)
13. **b** (You should know the common strong acids and bases. They are listed in Skill 3 in this chapter of the guide. Anything not on that list is likely to be weak.)
14. **d**
15. **c** (See the answer to Question 5 above. NH_4^+ would be acidic, the others neutral.)
16. **d** (There is no excess H^+ or OH^- to show being formed.)
17. **c** (HF and not H^+ is the major species in the solution of the weak acid. The base is strong, i.e., completely dissociated. Na^+ takes no part in the reaction.)
18. **c** (The weak base is simply NH_3; the strong acid is completely dissociated and so is represented simply as H^+.)
19. **b** (These are the proton donors.)
20. **c** (By definition of a Lewis acid.)

Solutions to Problems

1. (a) $pH = -Log[H^+] = 2.3$; $[H^+] = 10^{-2.3} = 5 \times 10^{-3}$
   (b) $[OH^-] = \dfrac{1.0 \times 10^{-14}}{5 \times 10^{-3}} = 2 \times 10^{-12}$
2. (a) $HI(g) \rightarrow H^+(aq) + I^-(aq)$
   (b) $HClO(aq) \rightleftharpoons H^+(aq) + ClO^-(aq)$
   (c) $F^-(aq) + H_2O \rightleftharpoons HF(aq) + OH^-(aq)$
   (d) $CO_3^{2-}(aq) + 2\ H^+(aq) \rightarrow H_2CO_3(aq)$ or $H_2O + CO_2(g)$
3. (a) $HNO_2(aq) + OH^-(aq) \rightarrow NO_2^-(aq) + H_2O$
   The solution formed contains NO_2^-, a basic anion. It is basic.
   (b) $V = \dfrac{0.156}{0.100} \times 25.0\ cm^3 = 39.0\ cm^3$
4. (a) NH_4^+, HSO_4^- can donate protons.
   (b) NH_3, SO_4^{2-}, HSO_4^- can accept protons.

5. (a) Since K_W increases with temperature ($10^{-13} > 10^{-14}$), ΔH must be positive.

   (b) A neutral solution at 60°C would have a pH of 6.5 . (Why?) Since this solution has a pH of 6.8, it must be basic.

SELECTED READINGS

Acid-base theory is discussed in a series of articles starting with:

Jensen, W.B., Lewis Acid-Base Theory, *Chemistry* (March 1974), pp. 11–14.

Some specific topics in acid-base chemistry:

Chilton, T.H., *Strong Water: Nitric Acid, Its Sources, Methods of Manufacture, and Uses,* Cambridge, Mass., MIT Press, 1968.
Giguere, P.A., The Great Fallacy of the H^+ Ion: And the True Nature of H_3O^+, *Journal of Chemical Education* (September 1979), pp. 571–575.
Likens, G.E., Acid Rain, *Scientific American* (October 1979), pp. 43–51.

Logarithms are discussed in Appendix 4 in the text.

18

IONIC EQUILIBRIA

QUESTIONS TO GUIDE YOUR STUDY

1. Can you predict the direction and extent of reaction for the dissolving of a slightly soluble ionic solid or its precipitation? For the reaction of an acid with a base?

2. Can Le Chatelier's Principle be applied to these equilibrium systems as for gaseous equilibria? For example, to predict the effect of a change in concentration of a reactant or product species?

3. How do you experimentally determine the equilibrium constant for the equilibrium system: slightly soluble solid $\rightleftharpoons$ solute ions? Or K for the dissociation of a weak acid or base?

4. What are some of the applications of solubility equilibria? weak acid equilibria? weak base equilibria?

5. What is the relation between the strength of an acid and the equilibrium constant for its dissociation? of a base and its constant?

6. What constitutes a buffer? How do you explain its properties?

7. How would you prepare a buffer solution, say one having a pH of 7?

8. How does one determine the concentration of H^+ in a water solution of a weak acid? The concentration of OH^- in a solution of a weak base?

9. How is the dissociation constant of a weak acid, HB, related to the constant for the conjugate weak base, B^-?

10. What is the relationship between the equilibrium constants for forward and reverse reactions?

YOU WILL NEED TO KNOW

This chapter relies very heavily on the concepts introduced in Chapters 16 and 17. The equilibrium constants considered here refer to reactions discussed in those chapters. The concepts listed below are particularly important.

Concepts

1. How to recognize an equation as representing the
    a. dissolving of a slightly soluble ionic solid; or, the reverse of such a reaction, the precipitation of a slightly soluble ionic solid. See Skill 4 in Chapter 16 of this guide.
    b. dissociation of a weak acid; or, the reverse reaction. See Skill 4 in Chapter 17.
    c. reaction of a weak base with water. See Skill 4 in Chapter 17.

Math

1. How to take roots on your calculator. In working problems involving weak acid equilibria and solubility equilibria, you will frequently be required to take square roots. In dealing with solubility equilibria, you will occasionally be required to take cube roots. Make sure you know how to do this.
2. How to take logarithms and antilogarithms (to find pH from $[H^+]$ or vice versa).
3. How to solve quadratic equations. Recall the discussion of the use of the quadratic formula in Chapter 13 of this guide. Example 18.7 introduces an approximation method that is very useful in solving the types of quadratic equations that arise in ionic equilibria. Also see the discussion of the "5% rule" directly following that example. A useful method of successive approximations is presented in Example 18.8.

BASIC SKILLS

Since this chapter deals with reactions and equations considered in Chapters 16 and 17, it may be useful at this point to quickly read over the Basic Skills of Chapter 16 (particularly 4 and 5) and 17 (especially Skills 1–4 and 6) in this guide. If you feel unsure of any of these skills, work through the related examples in the text, this time for yourself, and try some more problems. Then consider the following skills.

1. **Write the equilibrium constant expression:**
    a. K_{sp}, **for the dissolving of a slightly soluble ionic solid.**
    b. K_a, **for the dissociation of a weak acid in water solution.**
    c. K_b, **for the reaction of a weak base with water.**

This skill is fundamental to most of the examples and problems in this chapter. It is basically the same as Skill 1 in Chapter 13. Here, however, you are concerned with dissolved species. The expression should show the equilibrium concentration of any and all *dissolved* species.

Examples of each type of expression include:

$Ag_2 S(s) \rightleftharpoons 2\ Ag^+(aq) + S^{2-}(aq)$ $K_{sp} = [Ag^+]^2\ [S^{2-}]$

$HNO_2(aq) \rightleftharpoons H^+(aq) + NO_2^-(aq)$ $K_a = [H^+]\ [NO_2^-]/[HNO_2]$

$CO_3^{2-}(aq) + H_2 O \rightleftharpoons HCO_3^-(aq) + OH^-(aq)$ $K_b = [HCO_3^-]\ [OH^-]/[CO_3^{2-}]$

- -

Write the acid dissociation constant, K_a, for the weak acid HSO_3^-. _____ Write the base dissociation constant for SO_3^{2-}. _____

Begin by writing the equation for the reaction. Then write the expression for the equilibrium constant, placing product concentrations in the numerator. Exponents on these concentrations are the same as the coefficients in the balanced equations.

For the acid: $HSO_3^-(aq) \rightleftharpoons H^+(aq) + SO_3^{2-}(aq)$,

$K_a = [H^+][SO_3^{2-}]/[HSO_3^-]$

For the base: $SO_3^{2-}(aq) + H_2 O \rightleftharpoons HSO_3^-(aq) + OH^-(aq)$,

$K_b = [HSO_3^-][OH^-]/[SO_3^{2-}]$

- -

See the catalog of problems for further illustrations.

2. **Given the value of the equilibrium constant and all but one equilibrium concentration, calculate the remaining equilibrium concentration.**

Example 18.1 applies this skill to a slightly soluble solid, using K_{sp}. Consider the following example, using K_a.

- -

To a solution of a weak acid HB is added enough of the conjugate base to give $[B^-] = 2 \times [HB]$. If the value of K_a is known to be 1.0×10^{-4}, what must be $[H^+]$? _____

$K_a = 1.0 \times 10^{-4} = [H^+][B^-]/[HB]$ (What is the equation for the dissociation?) Solving for $[H^+]$:

$$[H^+] = 1.0 \times 10^{-4} \times \frac{[HB]}{[B^-]} = 1.0 \times 10^{-4} \times \frac{1}{2} = 5.0 \times 10^{-5}$$

This skill is much the same as 3a in Chapter 13.

3. Given the value of K_{sp}:
     a. predict whether or not a precipitate forms on mixing two solutions.

Example 18.2 illustrates this skill. The principle here is a simple one: a precipitate will form only if the product of concentrations, raised to the powers given by the expression for K_{sp}, exceeds the value of K_{sp}. Otherwise, the solubility equilibrium will not be established.

     b. calculate the solubility of the slightly soluble ionic solid. Perform the reverse calculation, K_{sp} from solubility.

Example 18.3 works the calculation in both directions. Note that the relationship between these two quantities depends on the type of formula, as shown in Table 18.2. Generally, you can qualitatively compare the solubilities of two solids by comparing their K_{sp} values *only* if they have the same type of formula.

4. Determine the value of K_a (or K_b) from the initial concentration of weak acid (base) and one equilibrium concentration.

Example 18.4 applies this skill to the water solution of a weak acid. Note the two fundamental relationships which must apply when the solution is formed simply by dissolving the acid HB in water:

$$[H^+] = [B^-] \text{ and } [HB] = \text{original conc. HB} - [H^+]$$

Calculate K_b for $NH_3(aq)$ if $[OH^-] = 1.9 \times 10^{-3}$ in a 0.20 M solution.

The reaction is $NH_3(aq) + H_2O \rightleftharpoons NH_4^+(aq) + OH^-(aq)$, for which K_b has the expression $K_b = [NH_4^+][OH^-]/[NH_3]$.

Note that in this solution all the OH^- and NH_4^+ come from the reaction written. (A negligible amount of OH^- comes from the dissociation of water itself.) So, $[OH^-] = [NH_4^+]$. Also, $[NH_3] = 0.20 - [OH^-]$.

Substituting in the expression for K_b:

$$K_b = (1.9 \times 10^{-3})(1.9 \times 10^{-3})/(0.20 - 1.9 \times 10^{-3})$$

$$= (1.9 \times 10^{-3})^2/0.20$$

$$= 1.8 \times 10^{-5}$$

- -

5. Use K_a (or K_b) to determine:
   a. $[H^+]$ in a buffer solution, given the initial concentrations of the conjugate acid-base pair.
   b. $[H^+]$ in a buffer, after adding a given quantity of strong acid or base.

Any calculation of this sort can be considered as consisting of two steps. First, the ratio $[HB]/[B^-]$ is determined. For the original buffer, the amounts of HB and B^- are either given or readily calculated from the statement of the problem.

Addition of x moles of a strong acid to the buffer increases the number of moles of HB by x and decreases the number of moles of B^- by the same amount. Conversely, if x moles of a strong base are added, the number of moles of B^- increases by x while that of HB decreases by x. One point to keep in mind: since HB and B^- are present in the same solution, their concentrations must be in the same ratio as the numbers of moles, i.e.,

$$\frac{[HB]}{[B^-]} = \frac{\text{no. moles HB}}{\text{no. moles B}^-}$$

Second, $[H^+]$ is calculated from the equation for K_a. That is,

$$[H^+] = K_a \times \frac{[HB]}{[B^-]}$$

Typical calculations are shown in Examples 18.5 and 18.6.

6. Use K_a (or K_b) to determine $[H^+]$ in a solution of a weak acid (base), given the initial concentration of the weak acid (base).

The calculation of $[H^+]$ is discussed in considerable detail in Examples 18.7 and 18.8. In the first example, the concentration of H^+ is so small compared to the original concentration of weak acid that the approximation

$$[HB] = \text{original conc. HB} - [H^+] \approx \text{original conc. HB}$$

is entirely justified. In contrast, in Example 18.8 the $[H^+]$ calculated is an appreciable fraction of the original concentration of weak acid, too large to be ignored. For that reason, it is desirable to refine the calculation, using a second approximation more nearly valid than the first.

Example 18.10 is similar to Example 18.7, except K_b is involved instead of K_a. Of course, in any of these calculations, you should be able to relate $[H^+]$, $[OH^-]$, and pH. (See Skill 2 in Chapter 17.)

7. **Calculate the equilibrium constant for a reaction, given:**
    a. **the equilibrium constant for the reverse reaction.**

The principle involved here is a simple one: when an equation is turned around, the expression for the new equilibrium constant is the reciprocal of the old one. If K' represents the constant for the reverse reaction, then the forward reaction has the constant: $K = 1/K'$. If you do not immediately recognize the kind of reaction represented by a given equation, chances are that you may recognize the reaction if it were turned around.

What is the value of the equilibrium constant for the reaction:

$$HC_2H_3O_2(aq) + OH^-(aq) \rightleftharpoons C_2H_3O_2^-(aq) + H_2O \text{ ?} \underline{\hspace{1cm}}$$

If you rewrote the equation backwards, you would see that a species $(C_2H_3O_2^-)$ reacts with water to form a basic solution (OH^-). Such a reaction has an equilibrium constant called K_b. So, the given reaction has a constant $K = 1/K_b(C_2H_3O_2^-)$.

Example 18.11 provides another illustration of this skill.

    b. **the equilibrium constants for two or more other, related reactions.**

Both Examples 18.9 and 18.12 apply this skill. The first example makes use of the general relationship between conjugate acid-base pairs:

$$K_a \times K_b = K_w = 1.0 \times 10^{-14}$$

The second example applies the **rule of multiple equilibria:** If a reaction can be expressed as the sum of two other reactions, K for the overall reaction is the product of the equilibrium constants for the individual reactions. That is if Reaction 3 = Reaction 1 + Reaction 2, then $K_3 = K_1 \times K_2$.

Choosing the conjugate acid and base pair, NH_4^+ and NH_3, show that K_a $\times$ $K_b = K_w$ follows from the consideration of a sum of reactions. _____

Consider the reactions of acid and base for which you know the constants and their expressions:

$$NH_4^+(aq) \rightleftharpoons NH_3(aq) + H^+(aq) \qquad K_a = [NH_3][H^+]/[NH_4^+]$$

$$NH_3(aq) + H_2O \rightleftharpoons NH_4^+(aq) + OH^-(aq) \quad K_b = [NH_4^+][OH^-]/[NH_3]$$

The sum of these reactions is, after cancelling NH_3 and NH_4^+ from both sides:

$$H_2O \rightleftharpoons H^+(aq) + OH^-(aq) \text{ for which } K = K_w$$

The multiplication of K_a by K_b does indeed give K_w. (Try it!)
So, the expression $K_a \times K_b = K_w$ is simply a specific example of the rule of multiple equilibria.

SELF-TEST

True or False

1. The solubility of AgCl in 0.10 M NaCl is greater than it is in   (   )
pure water.

2. Any two ionic solids that have the same K_{sp} value will have   (   )
the same solubility.

3. If (conc. M^+) $\times$ (conc. X^-) exceeds the value of K_{sp} of MX,   (   )
then a precipitate is expected to form.

4. HF ($K_a = 7 \times 10^{-4}$) is a stronger acid than HNO_2 ($K_a = 4 \times$ (   )
10^{-4}).

5. The equilibrium constant for the reaction of a strong acid   (   )
with a strong base is 1.0×10^{-14}.

6. A buffer would result if you mixed 100 cm^3 of 1 M HCl   (   )
with 100 cm^3 of 2 M $NaC_2H_3O_2$.

7. The pH of a buffer should not be affected by dilution with   (   )
water.

Multiple Choice

8. If $K_{sp} = 1 \times 10^{-12}$ for $PbCO_3$, then in a solution where ( )
$[CO_3{}^{2-}] = 0.2$, the equilibrium concentration of Pb^{2+} is (M):
    (a) 1×10^{-12} (b) 5×10^{-12}
    (c) 2×10^{-11} (d) 1×10^{-6}

9. The expression for K_{sp} for $As_2 S_3$ is: ( )
    (a) $[As_2{}^{3+}][S_3{}^{2-}]$ (b) $[As^{3+}]^2[S^{2-}]^3$
    (c) $[As^{3+}]^3[S^{2-}]^2$ (d) none of these

10. For an ionic solid of formula MX_2, the solubility s will be ( )
related to the K_{sp} by
    (a) $s = K_{sp}$ (b) $s^2 = K_{sp}$
    (c) $2 s^3 = K_{sp}$ (d) $4 s^3 = K_{sp}$

11. The addition of $AgNO_3$ to a saturated solution of AgCl ( )
would
    (a) cause more AgCl to precipitate
    (b) increase the solubility of AgCl due to the interionic
         attraction of $NO_3{}^-$ and Ag^+
    (c) lower the value of K_{sp} for AgCl
    (d) shift to the right the equilibrium $AgCl(s) \rightleftharpoons Ag^+(aq) +$
         $Cl^-(aq)$

12. Sodium chloride is soluble in water; yet, when concentrated ( )
hydrochloric acid is added to a saturated solution of this salt, NaCl(s)
precipitates out. Why?
    (a) HCl is a strong acid, and any strong acid will cause the
         precipitation
    (b) The common ion, Cl^-, shifts the equilibrium; NaCl(s)
         forms so that $[Na^+][Cl^-]$ remains constant
    (c) The K_{sp} decreases in the presence of acid
    (d) The K_{sp} is unaffected by acid, but is reduced by the
         increase in $[Cl^-]$.

13. $CaSO_4$ has a $K_{sp} = 3 \times 10^{-5}$. In which of the following ( )
should $CaSO_4$ be most soluble?
    (a) 1 M $K_2 SO_4$ (b) 2 M $CaCl_2$
    (c) pure $H_2 O$ (d) same in all three

14. K_b for NH_3 is 2×10^{-5}. K_a for the $NH_4{}^+$ ion is ( )
    (a) 2×10^{-5} (b) 5×10^{-9}
    (c) 5×10^{-10} (d) 2×10^{-19}

15. For the reaction of hydrochloric acid with sodium acetate ( )

$$H^+(aq) + C_2 H_3 O_2{}^-(aq) \rightleftharpoons HC_2 H_3 O_2 (aq)$$

the equilibrium constant is the same as
      (a) K_a for $HC_2H_3O_2$ (b) K_b for $C_2H_3O_2^-$
      (c) $1/K_a$ (d) K_w/K_b

16. If you had to choose between 0.10 M $HC_2H_3O_2$ ($K_a = 1.8$ (   )
$\times 10^{-5}$) and 0.10 M $HCHO_2$ ($K_a = 1.8 \times 10^{-4}$), which would have
the lower pH?
      (a) $HC_2H_3O_2$ (b) $HCHO_2$
      (c) both would be the same   (d) it would depend on
                                          the volume

17. A certain buffer contains equal concentrations of X^- and   (   )
HX. The K_a of HX is 10^{-10}. The pH of the buffer is
      (a) 4   (b) 7
      (c) 10   (d) 14

18. A buffer is formed by adding 500 cm^3 of 0.20 M $HC_2H_3O_2$ (   )
to 500 cm^3 of 0.10 M $NaC_2H_3O_2$. What is the maximum amount of
HCl that can be added to this solution without exceeding the
capacity of the buffer?
      (a) 0.01 mol   (b) 0.05 mol
      (c) 0.10 mol   (d) 0.20 mol

19. The pH of a 0.1 M solution of the weak acid HB is found to   (   )
be 3.0. What is the acid dissociation constant of HB?
      (a) 1×10^{-6} (b) 1×10^{-5}
      (c) 1×10^{-3} (d) 0.1

20. Consider the equilibrium system $NH_3(aq) + H_2O \rightleftharpoons$ (   )
$NH_4^+(aq) + OH^-(aq)$. The addition of a strong acid, such as HCl(aq),
will
      (a) increase $[H^+]$ (b) decrease $[NH_3]$
      (c) increase $[NH_4^+]$ (d) all the above

Problems

1. The solubility of $PbCl_2$ in water at a certain temperature is 1.6×10^{-2} M. Calculate:
      (a) K_{sp} of $PbCl_2$
      (b) the equilibrium concentration of Pb^{2+} when $PbCl_2$ (s) is added
            to 2.0 M NaCl

2. A bottle of the weak acid, aggravatic acid, labeled "0.040 M HA," is
found to have a pH of 2.70. Calculate:
      (a) $[H^+]$
      (b) $[HA]$
      (c) K_a for the acid

3. A buffer is prepared by mixing 0.10 mol of $HC_2H_3O_2$ ($K_a = 1.8 \times 10^{-5}$) with 0.10 mol of $C_2H_3O_2^-$.

    (a) What is the pH of the buffer?

    (b) What is the pH of the buffer after addition of 0.02 mol of NaOH?

4. The dissociation constants of the weak acids $HC_2H_3O_2$ and H_2CO_3 are 1.8×10^{-5} and 4.2×10^{-7} respectively. Calculate K for each of the following reactions ($K_w = 1.0 \times 10^{-14}$).

    (a) $C_2H_3O_2^-(aq) + H^+(aq) \rightarrow HC_2H_3O_2(aq)$

    (b) $HC_2H_3O_2(aq) + OH^-(aq) \rightarrow C_2H_3O_2^-(aq) + H_2O$

    (c) $H_2CO_3(aq) + C_2H_3O_2^-(aq) \rightarrow HCO_3^-(aq) + HC_2H_3O_2(aq)$

5. For HSO_4^-, $K_a = 1.2 \times 10^{-2}$. Using successive approximations, calculate $[H^+]$ in 1.0 M $NaHSO_4$.

SELF-TEST ANSWERS

1. **F** (The equilibrium $AgCl(s) \rightleftharpoons Ag^+(aq) + Cl^-(aq)$ would be shifted to the left by the common ion, Cl^-.)

2. **F** (Unless they have the same type formula. See Table 18.2.)

3. **T** (So that a state of equilibrium is approached. If this ion product were less than K_{sp}, then either or both concentrations could increase up to the value of the product = K_{sp}.)

4. **T** (They have the same type formula, HB. The larger K_a, the greater the extent of dissociation and the stronger the acid.)

5. **F** (If the only reaction carried out is neutralization, then K = $1/K_w = 1.0 \times 10^{14}$. Such a reaction is indeed "complete.")

6. **T** (Half the acetate would combine with H^+ to give $HC_2H_3O_2$. There would then be a mixture of this weak acid and its conjugate base.)

7. **T** (Both [HB] and $[B^-]$ would change in the same way. Their ratio would not.)

8. **b** ($[Pb^{2+}] = K_{sp}/[CO_3^{2-}]$.)

9. **b** (Write out the equation for the reaction; exponents are the coefficients.)

10. **d** (The concentration of M^{2+} is s; that of X^- is 2s. $(s)(2s)^2 = 4s^3$.)

11. **a** (The equilibrium in (d) is shifted to the *left*.)

12. **b** (Same reasoning as for Question 11.)

13. **c** (The common ions will decrease the solubility.)

14. **c** ($K_a = K_w/K_b$ for a conjugate acid-base pair.)

15. **c** (K_a is the equilibrium constant for the reverse reaction.)

16. b (Given the same type formula, HB, the stronger acid has the larger K_a. For the same initial concentration, HCHO gives the larger $[H^+]$ and the lower pH.)

17. c (Here, $K_a = [H^+] [X^-]/[HX] = 10^{-10} = [H^+]$. So pH = $-Log(10^{-10})$. The buffer is basic: X^- is stronger as a base, $K_b = 10^{-4}$, than HX is as an acid.)

18. b (Enough to react with all the weak base: moles $C_2 H_3 O_2^- = 0.50$. $\times 0.10 = 0.050$.)

19. b (Approximately, $K_a = [H^+]^2/0.1 = (10^{-3})^2/0.1 = 10^{-5}$. The approximations that $[H^+] = [B^-]$ and $[HB] = 0.1 - [H^+] = 0.1$ are both good. See Skill 6.)

20. d (The added H^+ will, in part, remove OH^- to form water and shift the equilibrium to the right. Or, you can consider the net result as one of shifting the equilibrium: $NH_3 (aq) + H^+(aq) \rightleftharpoons NH_4^+(aq)$.)

Solutions to Problems

1. The reaction is $PbCl_2 (s) \rightleftharpoons Pb^{2+}(aq) + 2 Cl^-(aq)$
    (a) $[Pb^{2+}] = s$, $[Cl^-] = 2 \times [Pb^{2+}] = 2s$
       $K_{sp} = [Pb^{2+}] [Cl^-]^2 = (s)(2s)^2 = 4s^3$
       $= 4(1.6 \times 10^{-2})^3 = 1.6 \times 10^{-5}$
    (b) $[Pb^{2+}] = K_{sp}/[Cl^-]^2 = 1.6 \times 10^{-5}/(2.0)^2 = 4.0 \times 10^{-6}$

2. The reaction is $HA(aq) \rightleftharpoons H^+(aq) + A^-(aq)$
    (a) $pH = -Log[H^+] = 2.70$
       $[H^+] = 10^{-2.70} = 2.0 \times 10^{-3}$
    (b) $[HA] = 0.040 - [H^+] = 0.040 - 0.0020 = 0.038$
    (c) $K_a = [H^+][A^-]/[HA] = (2.0 \times 10^{-3})^2/0.038 = 1.1 \times 10^{-4}$

3. (a) $[H^+] = 1.8 \times 10^{-5} \times \dfrac{[HC_2 H_3 O_2]}{[C_2 H_3 O_2^-]} = 1.8 \times 10^{-5} \times \dfrac{0.10}{0.10}$
       $= 1.8 \times 10^{-5}$
       $pH = 4.74$
    (b) $[H^+] = 1.8 \times 10^{-5} \times \dfrac{0.08}{0.12} = 1.2 \times 10^{-5}$; $pH = 4.92$

4. (a) Consider the reverse reaction. Its K is $K_a (HC_2 H_3 O_2)$.
       $K = 1/K_a = 1/(1.8 \times 10^{-5}) = 5.6 \times 10^4$
    (b) Again, consider the reverse reaction, for which
       $K = K_b (C_2 H_3 O_2^-)$
       $K = 1/K_b = K_a/K_w = (1.8 \times 10^{-5})/(1.0 \times 10^{-14}) = 1.8 \times 10^9$
    (c) Consider the sum of reactions:
       $H_2 CO_3 (aq) \rightarrow H^+(aq) + HCO_3^-(aq)$ $K_1 = K_a = 4.2 \times 10^{-7}$
       $C_2 H_3 O_2^-(aq) + H^+(aq) \rightarrow HC_2 H_3 O_2 (aq)$ $K_2 = 1/K_a = 5.6 \times 10^4$

The sum of these two gives the desired reaction for which K must be $K = K_1 \times K_2 = 2.4 \times 10^{-2}$.

5. 1st approx.: $[H^+]^2 \approx 1.0(1.2 \times 10^{-2})$; $[H^+] = 0.11$ M
    2nd approx.: $[H^+]^2 \approx 0.89(1.2 \times 10^{-2})$; $[H^+] = 0.10$ M

SELECTED READINGS

For additional practice in math and problem solving, see the readings listed in the Preface. Again, these readings also deal specifically with equilibrium calculations.

Acids and bases are considered in the articles by Jensen listed in Chapter 17 and, along with buffers:

Lott, J.A., Hydrogen Ions in Blood, *Chemistry* (May 1978), pp. 6–11.

Mogul, P.H., Dilute Solutions of Strong Acids: the Effect of Water on pH, *Chemistry* (October 1969), pp. 14–17.

Recall the application of Le Chatelier's Principle in Chapter 16:

Bodner, G.M., On the Misuse of Le Chatelier's Principle for the Prediction of the Temperature Dependence on the Solubility of Salts, *Journal of Chemical Education* (February 1980), pp. 117–119.

COORDINATION COMPOUNDS;
COMPLEX IONS

QUESTIONS TO GUIDE YOUR STUDY

1. What is a *complex ion?* A *coordination compound?* What special properties do they possess?

2. Where are you likely to encounter complex ions? What elements are most likely to form complexes?

3. What kind of experimental support is there for the existence of complexes? How could you show that they exist in the solid state, or in solution?

4. What is the nature of the bonding in these species? What geometries do you associate with various complexes?

5. How does the bonding and geometry of complexes account for their properties?

6. Why do complexes form? What energy effects accompany their formation and reactions?

7. How can you decide on the relative stabilities of complexes? How can you measure the extent to which one species is formed at the expense of another? How is this quantitatively expressed?

8. How can you change the extent of a reaction in which a complex is formed or decomposed?

9. When a complex ion takes part in a reaction, how does the rate depend on the nature of the complex?

10. What uses have been found for complexes? Can you justify their study? What natural products contain complex ions?

YOU WILL NEED TO KNOW

Concepts

1. The principle of electrical neutrality. This has been assumed in much of the discussion since it was first introduced in Chapter 2.

2. The formulas and charges of some common ions. See Table 3.2 and Figure 3.3 in the text.

3. The shapes and orientations of atomic orbitals. See Figure 19.12 as well as the figures of Chapter 8. The orientations of hybrid orbitals. See Chapter 10. (Review Skill 5 in Chapter 10 of this guide.)

4. How to write electron configurations and orbital diagrams. See Skills 4 and 5 in Chapter 8 of this guide.

5. How to write Lewis structures of simple molecules and ions. See Skill 2 in Chapter 10 of this guide.

Math

1. How to relate the wavelength of a spectral line to the change in energy of an atom. See Skill 1 in Chapter 8.

2. How to predict the effect of a change in concentration on the position of equilibrium. See Skill 5 in Chapter 13.

3. How to define the half-life of a reaction. See Skill 4 of Chapter 14.

BASIC SKILLS

1. **Given the formula of a complex ion or coordination compound, determine the charge on the central metal atom.**

Example 19.1 illustrates this skill. As seen in the discussion of this example, you need to recognize the charges and formulas of common ions. You also need to be familiar with several neutral molecular species, most of which you have seen many times already. These include H_2O, NH_3, and introduced in this chapter, en (ethylenediamine). Turning this skill around, you should be able to determine the overall charge on a complex if you are given the individual species which make it up. As for any other formula, the overall charge is simply the algebraic sum of the charges on all its parts.

Note that square brackets are used to set off the formula of a complex ion within the formula of a neutral compound. Recognizing, for example, that K is a Group 1 atom, therefore forms a +1 ion (see Chapter 9), tells you that the complex in $K_2[PtCl_6]$ carries a charge of -2.

2. **Given the formula of a complex, sketch its geometry, including any geometrical isomers.**

In essence, this skill involves three steps.

— *Determine the coordination number.* This is ordinarily easy. The species $Co(NH_3)_6{}^{3+}$, $Zn(NH_3)_4{}^{2+}$, and $Ag(NH_3)_2{}^{+}$ are seen to have coordination numbers of 6, 4, and 2, respectively. For example, the first of these involves Co^{3+} bonded to 6 nitrogen atoms. These atoms furnish the pairs of bonding electrons, one pair per nitrogen atom.

If one or more of the ligands is a chelating agent, the coordination number may not be so obvious. Consider, for example, the complexes $Co(NH_3)_4(en)^{3+}$, $Co(NH_3)_2(en)_2{}^{3+}$, and $Co(en)_3{}^{3+}$. Since each ethylenediamine molecule bonds to Co^{3+} at two different places, the coordination number in each of these species is 6.

— *Relate the coordination number to the geometry (apply electron pair repulsion ideas of Chapter 10).* If the coordination number is 2, the complex is linear; a coordination number of 6 corresponds to an octahedral complex. For coordination number 4, two geometries turn out to be possible: tetrahedral or square. There is no simple way to predict which geometry will prevail with a particular ion.

— *Having decided upon the geometry, sketch the complex.* This is entirely straightforward for linear and tetrahedral complexes. With square complexes you must keep in mind the possibility of geometrical isomerism in complexes of the form Ma_2b_2 and Ma_2bc, where a, b, and c are different ligands. Writing all of the isomers for an octahedral complex can be a little tougher. Example 19.2 suggests a logical approach to follow. Note in particular that for any given site in an octahedral complex, there is one site that is *trans* to it and four that are *cis*. This remains true regardless of which site you choose as your starting point.

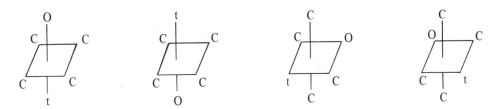

O = original site; C = *cis* position; t = *trans* position

3. **Write the electron configuration and orbital diagram for a given transition metal ion.**

Example 19.3 shows how this can be done. Note that in any of these *ions,* there are no outer s electrons. To write the configuration, you may as well fill the orbitals in order of increasing principal quantum number n (3p followed by 3d; 4p followed by 4d; etc.).

_ _

Write the configuration and, for the outermost (d) electrons, the orbital diagram for $_{27}Co^{3+}$. _____

In this ion there is a total of 27 − 3 or 24 electrons. This gives the argon core of 18 (filling through 3p) plus 6 more:

$$[_{18}Ar]\ 3d^6$$

3d

So the orbital diagram is: (↑↓) (↑) (↑) (↑) (↑)

— —

4. **Given the composition of a complex and its geometry, draw an orbital diagram for the electrons around the central metal atom and show the hybrid orbitals used.**

Several such diagrams are given in the text; the process used to obtain them is described there and illustrated in Example 19.4. Note that:

— the orbital diagram ordinarily includes only those electrons beyond the nearest noble gas core of electrons. In the structures on these pages, the 18 argon electrons are not shown.

— the number of orbitals filled by the bonding electrons, contributed by the ligands, is always equal to the coordination number, regardless of the nature of the ligands.

A systematic approach to drawing orbital diagrams might involve the four steps listed below. For simplicity, we assume here that the metal involved is in the first transition series (atomic number 21–30), but the same general approach can be extended to other metals. (Obtaining electron structures by analogy was illustrated in Chapter 8.)

a. Decide which orbitals are filled by the bonding electrons. This requires only that you know the geometry of the complex.

Coord. no.	Geometry	Hybridization	Orbitals filled (first transition series)
2	linear	sp	one 4s, one 4p
4	square	dsp^2	one 3d, one 4s, two 4p
4	tetrahedral	sp^3	one 4s, three 4p
6	octahedral	d^2sp^3	two 3d, one 4s, three 4p

b. Fill these orbitals with the bonding electrons and leave them there!

c. Decide how many electrons are contributed to the orbital diagram by the central metal ion. To do this, start with the atomic number of the metal, subtract those electrons lost in forming the ion, and finally take away the 18 Ar electrons.

d. If possible, fit the electrons left after step c into the 3d orbitals, following Hund's rule. Occasionally, you may find that there are too many electrons to fit into the available 3d orbitals. If this happens, put the overflow into the next highest empty orbital.

The application of this approach is shown in the following example.

_ _

Draw orbital diagrams for the square and tetrahedral complexes of Ni^{2+}.

We start by noting that in the square complex the hybridization is dsp^2, while in the tetrahedral complex it is sp^3 (step a). These orbitals are then filled by the bonding electrons (step b).

square

3d	4s	4p

$(\ )(\ )(\ )(\ )(\uparrow\downarrow)(\uparrow\downarrow)(\uparrow\downarrow)(\uparrow\downarrow)(\ )$

tetrahedral

3d	4s	4p

$(\ )(\ )(\ )(\ )(\ )(\uparrow\downarrow)(\uparrow\downarrow)(\uparrow\downarrow)(\uparrow\downarrow)$

Since the atomic number of Ni is 28, there are 28 electrons in the Ni atom. In the Ni^{2+} ion, there are $28 - 2 = 26 \ e^-$. Subtracting the 18 argon electrons gives us $26 - 18 = 8 \ e^-$ contributed to the orbital diagram by the Ni^{2+} ion (step c). These are just enough to fill all the available 3d orbitals in the square complex. In the tetrahedral complex, we distribute the 8 electrons among the five 3d orbitals so as to give the maximum number of unpaired electrons (step d).

square Ni^{2+}

3d	4s	4p

$(\uparrow\downarrow)(\uparrow\downarrow)(\uparrow\downarrow)(\uparrow\downarrow)(\uparrow\downarrow)(\uparrow\downarrow)(\uparrow\downarrow)(\uparrow\downarrow)(\ )$

tetrahedral Ni^{2+}

3d	4s	4p

$(\uparrow\downarrow)(\uparrow\downarrow)(\uparrow\downarrow)(\uparrow \ )(\uparrow \ )(\uparrow\downarrow)(\uparrow\downarrow)(\uparrow\downarrow)(\uparrow\downarrow)$

_ _

5. For any octahedral complex, draw orbital diagrams for the "high spin" and "low spin" forms (crystal field model).

The process involved here is illustrated in Example 19.5 Note that two steps are involved:

   a.   The five 3d orbitals are split into a lower energy group of three orbitals and a higher energy group of two orbitals.

   b.   The d electrons in the metal ion are located:

      — according to Hund's rule for the "high spin" complex, ignoring the energy difference between the two groups of orbitals.

      — preferentially in the lower energy orbitals for the "low spin" complex.

Note also that only if there are four to seven electrons to distribute can you distinguish high and low spin forms.

SELF-TEST

True or False

1. Of the two elements, Ca and Cu, the more likely to form more numerous and stable coordination compounds is Cu.   (   )

2. The *cis* isomer of $[Pt(NH_3)_2 Cl_2]$ is expected to be polar.   (   )

3. You would expect that $[Pt(NH_3)_4]Cl_2$ is more soluble in water than is $[Pt(NH_3)_2 Cl_2]$.   (   )

4. The strength of the coordinate covalent bond is comparable to that of the intermolecular forces (dispersion, dipole . . .).   (   )

5. Any complex ion with a coordination number of four for the central atom has a tetrahedral structure.   (   )

6. Geometric *(cis, trans)* isomerism is not observed in tetra- hedral complexes.   (   )

7. The color often associated with complexes is best explained by the valence bond (atomic orbital) model.   (   )

Multiple Choice

8. In the $NiCl_4{}^{2-}$ ion, the total number of electrons around the Ni, including bonding pairs, (A.N. Ni = 28) is   )
      (a)   26   (b)   28
      (c)   34   (d)   36

9. The hybridization of gold in $Au(NH_3)_2{}^+$ is   (   )
      (a) $d^2 sp^3$ (b) dsp^2
      (c) sp^3 (d) sp

10. The maximum number of possible geometric isomers for a   ( )
complex having dsp^2 hybridization would be
    (a) two   (b) three
    (c) four   (d) none

11. Geometric isomers would be expected for   ( )
    (a) $Zn(NH_3)_4{}^{2+}$ (b) $Zn(H_2O)_2(OH)_2$
    (c) $Co(NH_3)_3Cl_2Br$ (d) $Au(NH_3)_2{}^+$

12. Complex ions are held together by   ( )
    (a) marital bonds   (b) municipal bonds
    (c) coordinate covalent bonds (d) James bonds

13. Complex ions of coordination number six have a geometric   ( )
structure that is
    (a) linear   (b) square
    (c) tetrahedral   (d) octahedral

14. AgCl may be brought into water solution by the addition of   ( )
    (a) NaCl   (b) $AgNO_3$
    (c) H_2O (d) NH_3

15. The electronic structure for the central atom in $Co(en)_2Cl_2{}^+$ ( )
is

    (a) (↑↓)(↑↓)(↑↓)( ) |(↑↓) (↑↓) (↑↓)(↑↓)( )|
    (b) (↑↓)(↑↓)(↑↓) |(↑↓)(↑↓) (↑↓) (↑↓)(↑↓)(↑↓)|
    (c) (↑↓)(↑↓)(↑↓)(↑↓)(↑) |(↑↓) (↑↓)(↑↓)(↑↓)|
    (d) (↑↓)(↑)(↑)(↑)(↑) |(↑↓) (↑↓)(↑↓)(↑↓)|

16. The dissociation constant for the complex ion $Zn(NH_3)_4{}^{2+}$ ( )
is the equilibrium constant for the reaction represented by the
equation
    (a) $Zn^{2+}(aq) + 4\ NH_3(aq) \rightleftharpoons Zn(NH_3)_4{}^{2+}(aq)$
    (b) $Zn(NH_3)_4{}^{2+}(aq) + H_2O \rightleftharpoons Zn(NH_3)_3(H_2O)^{2+}(aq) +$
        $NH_3(aq)$
    (c) $Zn(NH_3)_4{}^{2+} + 2\ e^- \rightleftharpoons Zn(s) + 4\ NH_3(aq)$
    (d) $Zn(NH_3)_4{}^{2+}(aq) \rightleftharpoons Zn^{2+}(aq) + 4\ NH_3(aq)$

17. Of the following solutions, all at the same initial concen-   ( )
tration, which would have the highest total concentration of solute
particles?
    (a) NaCl   (b) $CuCl_2$
    (c) $HC_2H_3O_2$ (d) $[Co(NH_3)_6]Cl_3$

18. How might you show experimentally that a particular   ( )
complex is square?
    (a) two isomers might be isolated; they would show
        different lability

(b) x-ray diffraction for the complex in the solid state would reveal the geometry

(c) magnetic measurements might distinguish between dsp^2 and sp^3 orbitals

(d) all of the above

19. The central metal atom in $[Co(NH_3)_5Cl](NO_3)_2$ can be ( ) assigned a charge of

(a) +1   (b) +2

(c) +3   (d) something else

20. The hybrid orbitals used by $_{13}Al$ in the complex $AlCl_4^-$ are ·( ) expected to be

(a) sp   (b) sp^3

(c) dsp^2 (d) no hybrids used

Problems

1. For each of the following compounds, give the charge on the complex as well as on the central metal atom in the complex.

(a) $Na[AgCl_2]$ (b) $[Ag(NH_3)_2]Cl$

(c) $[Zn(NH_3)_2(OH)_2]$ (d) $[Cr(en)_2(ox)]NO_3$

2. For each of the species listed in Problem 1, give the geometry of the complex ion and the hybrid orbitals used by the central metal atom.

3. A certain coordination compound has the formula $H_4[Co(CN)_6]$. Answer the following questions about the complex ion in this compound.

(a) What is the coordination number of the central atom?

(b) Write the orbital diagram of the metal atom in the complex ion, showing the hybrid orbitals occupied by the ligand electrons. (Atomic number Co = 27.)

(c) If the complex ion is of the "low spin" form, write its orbital diagram according to the crystal field model.

4. The following four coordination compounds of Cr^{3+} have been isolated. All have three unpaired d electrons. (Atomic number Cr = 24.)

	(a)	(b)	(c)	(d)
Formula	$Cr(NH_3)_3Cl_3$	$Cr(NH_3)_4Cl_3$	$Cr(NH_3)_4Cl_3$	$Cr(NH_3)_6Cl_3$
Color	green	brown	violet	yellow
Electrical conductivity	like CH_3OH	like NaCl	like NaCl	like $Al(NO_3)_3$

Sketch a plausible structure for each complex present.

5. Given K_{sp} $Zn(OH)_2 = 5 \times 10^{-17}$, K_d $Zn(NH_3)_4{}^{2+} = 1 \times 10^{-9}$

(a) Calculate K for the reaction:

$$Zn(OH)_2(s) + 4 NH_3(aq) \rightleftharpoons Zn(NH_3)_4{}^{2+}(aq) + 2 OH^-(aq)$$

(b) Determine the equilibrium concentration of $Zn(NH_3)_4{}^{2+}$ in a solution 0.10 M in NH_3 with a pH of 9.0.

SELF-TEST ANSWERS

1. **T** (Cu is a transition metal, Ca is not.)
2. **T** (Two Cl atoms at one "end" of the square complex, two NH_3 units at the other. Polar bonds would not cancel. Apply Skill 4 of Chapter 10.)
3. **T** (The first gives ions in solution; the second is a molecular species.)
4. **F** (Many complexes are very stable. Note that the valence bond model treats these bonds as covalent.)
5. **F** (Some are square planar complexes.)
6. **T** (Here is one way of distinguishing experimentally between square and tetrahedral complexes.)
7. **F** (Crystal field model.)
8. **c** (For Ni^{2+} and four pairs of electrons from the four Cl^- ions.)
9. **d** (Two hybrid orbitals are needed for the two N ligand atoms.)
10. **a** (Such hybrids give square planar geometry. So, *cis* and *trans* isomers.)
11. **c** (The zinc complexes are tetrahedral, as you might expect knowing that Zn^{2+} has no d orbitals available for forming dsp^2 hybrids.)
12. **c**
13. **d** (With six corners.)
14. **d** (To form a complex ion.)
15. **b** (en is a chelating agent with two pairs of electrons to donate.)
16. **d**
17. **d** (Giving 4 mol ions per mole of compound dissolving.)
18. **d** (Choices a and c are described or implied in Chapter 19; x-ray diffraction determines the relative positions of atoms.)
19. **c** (The complex itself carries a charge of +2; taking chlorine to be -1 leaves Co^{3+}.)
20. **b** (Al^{3+} has 3s and 3p orbitals available.)

Solutions to Problems

1. Charge on complex   Charge on metal

	Charge on complex	Charge on metal	
(a)	−1	+1	(Na is +1, Cl is −1)
(b)	+1	+1	(Cl is −1)
(c)	0	+2	(OH is −1)
(d)	+1	+3	(NO_3 is −1, ox is −2)

2. Geometry of complex   Hybrid orbitals
   - (a) linear   sp   (only two ligand atoms)
   - (b) linear   sp   (only two N ligand atoms)
   - (c) tetrahedral sp^3 (Zn^{2+} is d^{10}, so you might expect ligand electrons to start filling 4s, then 4p)

   (If you predict square, then the hybrids must be dsp^2.)
   - (d) octahedral $d^2 sp^3$ (en and ox are bidentate)

3. (a) 6
   - (b) Co^{2+} has 27 − 2 or 25 electrons; so it has the argon core of 18 plus 7 more. The ligands contribute 6 pairs.

   3d   4s   4p

   (↑↓) (↑↓) (↑↓) (↑↓) (↑↓)   (↑↓)   (↑↓) (↑↓) (↑↓)
                 4d

   (↑) ( ) ( ) ( ) ( )
   - (c) "Low spin" implies maximum pairing of electrons, so a large splitting. The metal electrons pile up in the lower energy group of d orbitals.   (↑) ( )

                                     (↑↓) (↑↓) (↑↓)

4. (a)   (b)

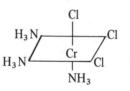 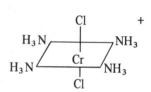

(c)   (d)

Note that the structures of b and c are interchangeable unless additional information is given. Also, another isomer is possible for a.

5. (a) Applying the rule of multiple equilibria (Chapter 18):

   $K = K_{sp}/K_d = (5 \times 10^{-17})/(1 \times 10^{-9}) = 5 \times 10^{-8}$
   - (b) $K = \dfrac{[Zn(NH_3)_4{}^{2+}] \times [OH^-]^2}{[NH_3]^4}$; $[OH^-] = \dfrac{1 \times 10^{-14}}{1 \times 10^{-9}} = 1 \times 10^{-5}$

   $[Zn(NH_3)_4{}^{2+}] = K \times \dfrac{[NH_3]^4}{[OH^-]^2} = \dfrac{(5 \times 10^{-8}) \times (1 \times 10^{-4})}{1 \times 10^{-10}}$

   $= 0.05 \text{ M}$

SELECTED READINGS

Alternative discussions, including more advanced treatment of bonding, can be found in the readings of Chapter 10 and:

Basolo, F., *Coordination Chemistry,* New York, W.A. Benjamin, 1964.

Cotton, F.A., Ligand Field Theory, *Journal of Chemical Education* (September 1964), pp. 466–476.

From reaction mechanisms to biological and medicinal complexes:

Bailar, J.C., Jr., Some Coordination Compounds in Biochemistry, *American Scientist* (September-October 1971), pp. 586–592.

House, J.E., Jr., Substitution Reactions in Metal Complexes, *Chemistry* (June 1970), pp. 11–14.

Jones, M.M., Therapeutic Chelating Agents, *Journal of Chemical Education* (June 1976), pp. 342–347.

Orna, M.V., Chemistry and Artists' Colors, *Journal of Chemical Education* (April 1980), pp. 264–267.

Perutz, M.F., Hemoglobin Structure and Respiratory Transport, *Scientific American* (December 1978), pp. 92–125.

Perutz, M.F., The Hemoglobin Molecule, *Scientific American* (November 1964), pp. 64–76.

Schubert, J., Chelation in Medicine, *Scientific American* (May 1968), pp. 40–50.

20

QUALITATIVE ANALYSIS

QUESTIONS TO GUIDE YOUR STUDY

1. What kinds of problems do you try to solve in qualitative analysis? How is this different from quantitative analysis?

2. To what kinds of systems (chemical, biological, mineralogical, . . .) can these procedures be applied?

3. What is the nature of the dissolved species encountered in the analyses? Are they simple molecules, ions, or complexes?

4. What properties allow the separation of cations into distinct groups? What properties allow the further separation and identification of these ions?

5. Specifically what conditions (temperature, concentration, pH, . . .) are controlled or varied to achieve these separations? What conditions favor the formation of one species instead of another; for example, AgCl(s) instead of $Ag(NH_3)_2^+(aq)$?

6. How are changes in any of these reaction conditions expected to affect the rate and extent of the reactions involved?

7. What reagents are commonly used? What kinds of reactions occur in the analyses? Can simple equations be written for each reaction?

8. Are the separations in fact complete? Are there interferences you need to consider? Can a simple test be performed which would allow you to confirm the presence of a given ion, or to distinguish between two given species?

9. Given the results of an analysis, exactly what can you say about the composition of an unknown?

10. What are the potential hazards of the procedures and materials used in the experiments of qualitative analysis? What safety precautions should you take? How do you dispose of the chemical wastes?

YOU WILL NEED TO KNOW

Concepts

    1. How to write balanced net ionic equations involving
       – precipitates, forming or dissolving. See Skill 4 in Chapter 16.
       – acids and bases. See Skills 4 and 6 in Chapter 17 of this guide.
       – complex ions. See, for example, Section 19.5 in the text.
    2. How, therefore, to predict the formulas of ionic compounds by applying the principle of electrical neutrality. How to recognize the formulas and charges of the common ions. See Table 3.2 and Figure 3.3 in the text.
    3. The solubility rules. See Table 16.3 in the text.

Math

    1. How to write the expression for the equilibrium constant for any reaction. This will include recognizing and using K_{sp}, K_w, K_a, K_b, and K_d.
    2. How to calculate the equilibrium constant for any given reaction, using the rule of multiple equilibria and the relation between the constants for forward and reverse reactions. See Skill 7 in Chapter 18.
    3. How to use any given equilibrium constant to determine whether or not a reaction should occur under given conditions. In particular, see Skill 3 in Chapter 18 of the guide.

BASIC SKILLS

    The majority of the skills used in this chapter have already been discussed. They are used throughout qualitative analysis. You should become thoroughly familiar with these skills, if you are not already. They include the following.
    – *Write the net ionic equation for the formation of a precipitate.*
    In the simplest case, a precipitate forms when solutions containing two different ions are mixed. Example 20.1a illustrates this. You first saw this skill as 4b in Chapter 16. Here, too, you need to know the solubility rules (Table 16.3) in order to predict the possible products.
    Sometimes, a molecular solute in water solution produces one of the ions required for a precipitate. An example is H_2S, which forms S^{2-} ions. Hydrogen sulfide precipitates the sulfides of Group II, which are very insoluble:

$$Cu^{2+}(aq) + H_2S(aq) \rightarrow CuS(s) + 2\,H^+(aq)$$

See the second part of Example 20.1. Another example of a molecular precipitating agent is ammonia. In water solution, NH_3 forms OH^- ions. Insoluble hydroxides, such as $Mn(OH)_2$ and $Zn(OH)_2$, can be precipitated by adding NH_3:

$$Zn^{2+}(aq) + 2\ NH_3(aq) + 2\ H_2O \rightarrow Zn(OH)_2(s) + 2\ NH_4^+(aq)$$

See Example 20.8. Equation 20.13 in the text shows how such an overall reaction may be considered in terms of simpler steps (one of them being the formation of OH^- by reaction of NH_3 with H_2O).

— *Write net ionic equations for the dissolving of precipitates.* This skill is illustrated in the body of the text, particularly in the last section of the chapter. Consider the following three classes of reactions.

  a. Strong acids (H^+) are effective in dissolving many water-insoluble compounds which contain a basic anion. In particular, a strong acid such as dilute (6 M) HCl will dissolve:
  — all insoluble hydroxides. A typical example is:

$$Zn(OH)_2(s) + 2\ H^+(aq) \rightarrow Zn^{2+}(aq) + 2\ H_2O$$

  — all insoluble carbonates.

$$ZnCO_3(s) + 2\ H^+(aq) \rightarrow Zn^{2+}(aq) + CO_2(g) + H_2O$$

  — some, but not all, insoluble sulfides.

$$ZnS(s) + 2\ H^+(aq) \rightarrow Zn^{2+}(aq) + H_2S(aq)$$

  b. Ammonia dissolves many insoluble compounds of Ag^+, Cu^{2+}, Cd^{2+}, Ni^{2+}, and Zn^{2+}. It does so by forming complex ions. See Table 20.2 in the text. A typical example includes:

$$Zn(OH)_2(s) + 4\ NH_3(aq) \rightarrow Zn(NH_3)_4^{2+}(aq) + 2\ OH^-(aq)$$

  c. Excess dilute (6 M) NaOH dissolves many insoluble compounds of Pb^{2+}, Sn^{2+}, Sb^{3+}, Al^{3+}, and Zn^{2+}. It does so by forming complex ions. See Table 20.2 in the text. An example:

$$Zn(OH)_2(s) + 2\ OH^-(aq) \rightarrow Zn(OH)_4^{2-}(aq)$$

— *Write the expression for the equilibrium constant for a given reaction.* Example 20.5 calls for this skill as applied to the dissociation of a complex ion. The general principle behind this skill has been seen in Skills 1 and 7 in Chapter 18 and in Section 19.5 in the text.

– Given the value of K_{sp},

a. and one equilibrium concentration, calculate the other equilibrium concentration. This is Skill 2 of Chapter 18. It is applied here in Example 20.7 in a straightforward manner. This example is similar to Example 18.1.

b. predict whether or not a precipitate forms on mixing two solutions. This is Skill 3a of Chapter 18. As before, you need to compare the ion product to the K_{sp} in order to decide if precipitation should occur. See Example 20.9 for an illustration of this skill.

– Calculate the equilibrium constant for a given reaction. See Skill 7 (parts a and b) in Chapter 18. Two important rules which are very helpful here were introduced in Chapter 18. They are the reciprocal relationship between the constants for forward and reverse reactions, and the rule of multiple equilibria. Example 20.10 illustrates the use of this skill in this chapter of the text.

Skills new to this chapter include the following.

1. Given the dissociation constant, K_d, for a complex ion, relate the concentration of aquated ("free") metal ion, complex ion, and ligand.

A typical calculation is shown in Example 20.5 No new principle is involved here. The constant K_d is handled in exactly the same way as the constants K_a and K_b (Chapter 18). What the constant means, and a table of K_d values, was given in Section 19.5 of the text.

2. Write the net ionic equation for the decomposition of a complex ion.

Example 20.6 illustrates this skill. The principle involved here is that complexes containing NH_3 or OH^- are stable in basic solution but decompose to give the "free" ion in strong acid.

- -

Write equations to show: (a) the decomposition of $Cu(NH_3)_4{}^{2+}$ in $HCl(aq)$; (b) the reaction of hydrochloric acid with a solution containing $Pb(OH)_3{}^-$.

In the first case, $H^+(aq)$ from the strong acid reacts with the weak base NH_3 of the complex ion. You might consider the reaction to proceed in steps:

$$Cu(NH_3)_4{}^{2+}(aq) \rightarrow Cu^{2+}(aq) + 4\,NH_3(aq)$$

$$4\,NH_3(aq) + 4\,H^+(aq) \rightarrow 4\,NH_4{}^+(aq)$$

overall: $Cu(NH_3)_4{}^{2+}(aq) + 4\,H^+(aq) \rightarrow Cu^{2+}(aq) + 4\,NH_4{}^+(aq)$

The reaction proceeds because K for the second step, $(1/K_a)^4$, is very large.

For the second reaction, $H^+(aq)$ reacts with the base OH^- of the complex ion. Once $Pb^{2+}(aq)$ is made available, it combines with $Cl^-(aq)$ to give insoluble $PbCl_2(s)$. Stepwise:

$$Pb(OH)_3^-(aq) \rightarrow Pb^{2+}(aq) + 3\ OH^-(aq)$$

$$3\ OH^-(aq) + 3\ H^+(aq) \rightarrow 3\ H_2O$$

$$\underline{Pb^{2+}(aq) + 2\ Cl^-(aq) \rightarrow PbCl_2(s)\hspace{3cm}}$$

so overall:

$$Pb(OH)_3^-(aq) + 3\ H^+(aq) + 2\ Cl^-(aq) \rightarrow PbCl_2(s) + 3\ H_2O$$

— —

3. **Devise an abbreviated scheme of analysis for an unknown containing a few selected ions.**

The ions involved may be from a single group or from more than one group. The simplest approach might be to start with Table 20.1, detailing the separation of cations into groups, and modify it by eliminating steps that apply to ions not present in the unknown. Likewise, for two or more ions present in the same group, it would be useful to start with the flow chart for that group and modify it. Example 20.2 illustrates this skill.

Again, in some problems, it may be effective to devise a new scheme consisting of only a few steps. To do this, you must be thoroughly familiar with the chemistry of the ions involved. Perhaps most useful here is a knowledge of Tables 16.3 (solubility rules), 20.1, and 20.2 (complexes of the cations of Groups I–IV with NH_3 and OH^-).

4. **Given the results of an analysis, state which ions in an unknown solution are present, absent, or in doubt.**

Examples 20.3 and 20.4 apply this skill. Here too you must know the information summarized in the tables referred to above (Skill 3). Some points you should keep in mind:

— An ion is *present* if it is required to explain a single positive test in the analysis.

— An ion is *absent* if a test for it is negative.

— An ion is *in doubt* if the description of the analysis does not apply to it. Again, if the results could be explained equally well by two or more different ions, those ions are in doubt.

— Each piece of information given must be consistent with the ions you list as present, absent, or in doubt. There can be no unexplained observations.

SELF-TEST

True or False

1. All chlorides are soluble except those of analytical Group I ( )
cations.

2. The mercury(I) cation is represented by Hg^+. ( )

3. All metal hydroxides are soluble in strong acid. ( )

4. A strong acid is any acid present at high concentration. ( )

5. An amphoteric hydroxide dissolves in either strong acid or ( )
strong base.

6. In a solution of H_2S, $[S^{2-}]$ increases as the pH is increased. ( )

Multiple Choice

7. Depending on its composition, a precipitate may turn out to ( )
be soluble in
    (a) hot water   (b) a dilute strong acid
    (c) a complexing agent   (d) any of these

8. What effect(s) may 6 M NaOH have on an aqueous solution? ( )
    (a) increase $[OH^-]$ (b) form a precipitate
    (c) form a complex ion   (d) any of these

9. The precipitation of silver chloride in Group I by the ( )
addition of hydrochloric acid is best represented by the equation
    (a) $Ag^+(aq) + HCl(aq) \rightarrow AgCl(s) + H^+(aq)$
    (b) $Ag^+(aq) + Cl^-(aq) \rightarrow AgCl(s)$
    (c) $Ag(NH_3)_2^+(aq) + Cl^- + 2\,H^+(aq) \rightarrow AgCl(s) +$
       $2\,NH_4^+(aq)$
    (d) some other equation

10. When $NH_3(aq)$ is added to a solution containing Zn^{2+}, a ( )
precipitate appears. Further addition of NH_3 dissolves the precipitate. The precipitate is
    (a) Zn   (b) $Zn(OH)_2$
    (c) $Zn(NH_3)_2$ (d) $Zn(NH_3)_4^{2+}$

11. The chloride from a Group I unknown is found to be ( )
entirely soluble in hot water. One must conclude that
    (a) Pb^{2+} is present   (b) Hg_2^{2+} is absent
    (c) Ag^+ is absent   (d) all of these

12. The chloride from a Group I unknown is found to be partly   (   )
soluble in NH_3 (aq). One can conclude that
     (a) Pb^{2+} is present   (b) Hg_2^{2+} is present
     (c) Ag^+ is present   (d) all of these

13. In a solution of pH 0.5, as in Group II analysis, $[H^+]$ (   )
     (a) is the same as $[H_2S]$ (b) is about 0.1 M
     (c) is about 0.3 M   (d) is about 0.5 M

14. The precipitation of $Al(OH)_3$ in Group III is done in the   (   )
presence of NH_4Cl. The NH_4^+
     (a) furnishes the needed OH^-
     (b) buffers the system
     (c) lowers the pH to below 7
     (d) complexes the Al^{3+}

15. Without looking them up, place each of the following ions   (   )
in the appropriate analytical group: Bi^{3+}, Co^{2+}, Hg^{2+}, Hg_2^{2+}, Mg^{2+},
Mn^{2+}, Pb^{2+}, and Zn^{2+}.
     (a) Group I _____   (b) Group II _____
     (c) Group III _____   (d) Group IV _____

16. Which of the following precipitates is colored (other than   (   )
white)?
     (a) $PbCl_2$ (b) $PbCrO_4$
     (c) $AgCl$ (d) $HgNH_2Cl$

17. Which cation forms stable complexes with both NH_3 and   (   )
OH^-?
     (a) Ag^+ (b) Cu^{2+}
     (c) Zn^{2+} (d) Hg_2^{2+}

18. A flame test may be used to confirm the presence of   (   )
     (a) Ag^+ (b) Al^{3+}
     (c) NH_4^+ (d) K^+

19. The addition of NH_3 (aq) to a solution of a metal ion may   (   )
result in
     (a) the formation of a precipitate containing OH^-
     (b) the formation of a complex ion containing NH_3
     (c) an increase in the pH
     (d) any of these

20. If a cation is known to form both an insoluble hydroxide   (   )
and a stable hydroxo complex, you should add _____ if you want
to precipitate the hydroxide.
     (a) excess $NaOH$(aq)   (b) NH_3 (aq)
     (c) NH_4^+(aq)   (d) HOH

Problems

1. Write equations for the reactions occurring when
   (a) the presence of lead in Group I is confirmed.
   (b) dilute HNO_3 is added to a solution formed by the addition of excess NH_3 to $AgCl(s)$.
   (c) excess $NH_3(aq)$ is added to a solution containing Cu^{2+}. (Two reactions occur.)
   (d) Bi_2S_3 is precipitated from Group II.

2. Using only the solubility rules,
   (a) design a scheme that would allow the separation of Cu^{2+} from a mixture containing only the cations Ag^+, Cu^{2+}, and Co^{2+}.
   (b) Construct the flow chart for such a separation.

3. In Group III analysis, a buffer of pH 9.0 is used. How many moles of NH_4Cl should you add to 10 cm^3 of a solution that is 1 M in NH_3 to arrive at such a pH? K_a $(NH_4^+) = 5.6 \times 10^{-10}$.

4. Calculate the equilibrium constant for the reaction:

$$Zn(OH)_2 (s) + 2 H^+(aq) \rightarrow Zn^{2+}(aq) + 2 H_2O$$

given the constants: K_{sp} of $Zn(OH)_2$ = 5×10^{-17}; $K_w = 1.0 \times 10^{-14}$.

5. A solution which may contain Pb^{2+}, Cd^{2+}, and Zn^{2+} is treated with excess $NaOH(aq)$. The precipitate which at first forms is finally entirely dissolved. What species are definitely present or absent, or in doubt?

SELF-TEST ANSWERS

1. **T** (You should *know* the solubility rules, Table 16.3, since they are heavily relied on in qualitative analysis.)
2. **F** (The actual species is diatomic, Hg_2^{2+}.)
3. **T** (K for such a reaction is related to $1/K_w$, a very large number. See Problem 4, below.)
4. **F** (Strength refers to the extent of dissociation. See Chapter 17.)
5. **T** (Examples include $Zn(OH)_2$ and $Al(OH)_3$. See Section 20.3 in the text.)
6. **T** (As pH goes up, $[H^+]$ goes down. So, $H_2S(aq) \rightleftharpoons 2 H^+(aq) + S^{2-}(aq)$ is shifted to the right. Apply Le Chatelier's principle.)
7. **d** (Some examples: (a) $PbCl_2$; (b) $Al(OH)_3$; (c) $AgCl$.)
8. **d** (For example; $Zn(OH)_2$, a precipitate, would dissolve on adding excess OH^- to give a complex ion.)

9. b (The strong acid is $H^+(aq)$ plus $Cl^-(aq)$. Choice c represents the confirmation of Ag^+ by the addition of another strong acid, say HNO_3, in the presence of Cl^-.)

10. b

11. d (Each of these conclusions is consistent with and required by the observation.)

12. c (Either or both Pb^{2+} and Hg_2^{2+} are also present.)

13. c ($pH = -Log[H^+] = 0.5; [H^+] = 10^{-0.5}$.)

14. b (And so prevents the formation of the stable complex ion $Al(OH)_4^-$. See Table 20.2.)

15. a: Hg_2^{2+}, Pb^{2+} b: Bi^{3+}, Hg^{2+} c: Co^{2+}, Mn^{2+}, Zn^{2+} d: Mg^{2+} (See Table 20.1.)

16. b (Yellow.)

17. c (The only one in Groups I–IV. See Table 20.2.)

18. d (Giving a violet flame.)

19. d (Example: see Problem 1c below.)

20. b (Excess NaOH would give the complex ion.)

Solutions to Problems

1. (a) $Pb^{2+}(aq) + CrO_4^{2-}(aq) \rightarrow PbCrO_4(s)$

   (b) $Ag(NH_3)_2^+(aq) + Cl^-(aq) + 2 H^+(aq) \rightarrow AgCl(s) + 2 NH_4^+(aq)$

   (c) See Equation 20.8 in the text.
   $Cu^{2+}(aq) + 2 NH_3(aq) + 2 H_2O \rightarrow Cu(OH)_2(s) + 2 NH_4^+(aq)$
   $Cu(OH)_2(s) + 4 NH_3(aq) \rightarrow Cu(NH_3)_4^{2+}(aq) + 2 OH^-(aq)$

   (d) $2 Bi^{3+}(aq) + 3 H_2S(aq) \rightarrow Bi_2S_3(s) + 6 H^+(aq)$
   This is Example 20.1b in the text.

2. (a) Consider the first two steps given in Table 20.1 in the text: add 6 M HCl to precipitate AgCl; to the solution, brought to a pH of 0.5, add H_2S to precipitate CuS. Discard the remaining solution (Co^{2+}).

   (b)

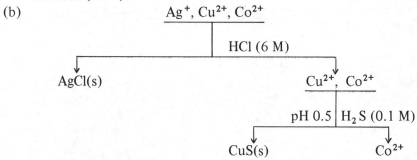

3. $[NH_4^+] = [NH_3][H^+]/K_a = (1)(1 \times 10^{-9})/(5.6 \times 10^{-10}) = 1.8$
   Needed: 1.8 M or 1.8×10^{-2} mol/10 cm^3

4. Consider the sum of reactions that gives the desired reaction:

$$Zn(OH)_2 (s) \rightarrow Zn^{2+}(aq) + 2\ OH^-(aq) \qquad K_1 = K_{sp}$$
$$2\ H^+(aq) + 2\ OH^-(aq) \rightarrow 2\ H_2O \qquad K_2 = 1/K_w^2$$

K for the sum is $K_1 \times K_2 = 5 \times 10^{11}$

So, $Zn(OH)_2$ should dissolve, like other metal hydroxides, in a strong acid.

5. Either Pb^{2+} or Zn^{2+}, or both, must be present. Cd^{2+} is definitely absent. See Table 20.2.

SELECTED READINGS

Alternative treatments of qualitative analysis, including different and more extensive schemes:

Feigl, F., *Spot Tests in Inorganic Analysis,* New York, American Elsevier, 1962.

Layde, D.C., *Introduction to Qualitative Analysis,* Boston, Allyn and Bacon, 1971.

Masterton, W.L., and Slowinski, E.J., *Chemical Principles with Qualitative Analysis,* Philadelphia, W.B. Saunders, 1978.

Slowinski, E.J., and Masterton, W.L., *Qualitative Analysis and the Properties of Ions in Aqueous Solution,* Philadelphia, W.B. Saunders, 1971.

Sorum, C., *Introduction to Semimicro Qualitative Analysis,* Englewood Cliffs, N.J., Prentice-Hall, 1967.

Stock, J.T., *Small-Scale Inorganic Qualitative Analysis,* New York, Chemical Publishing, 1963.

General and very useful references include the handbooks:

Handbook of Chemistry and Physics, Cleveland, CRC Press.

Lange's Handbook of Chemistry, New York, McGraw-Hill.

Some interesting and sometimes relevant descriptive chemistry:

Carter, K.N., Qualitative Detection of Manganese and Lead in Exhaust Deposits, *Journal of Chemical Education* (January 1978), p. 64.

Lingane, J.J., *Analytical Chemistry of Selected Metallic Elements,* New York, Reinhold, 1966.

McCreery, H.I., Sodium, *Chemistry* (September 1976), pp. 11–13.

Naik, D.V., Qualitative Analysis of Some Inorganic Anions by Paper Chromatography, *Chemistry* (May 1977), pp. 27–28.

Rabenstein, D.L., The Chemistry of Methylmercury Toxicology, *Journal of Chemical Education* (May 1978), pp. 292–296.

Scott, A.F., A Simple Test for Mercury, *Chemistry* (May 1975), pp. 29–30.

Treptow, R.S., Amalgam Dental Fillings, *Chemistry* (April 1978), pp. 15–19.

21

REDOX REACTIONS; STANDARD
VOLTAGES

QUESTIONS TO GUIDE YOUR STUDY

1. What reactions have you encountered that can be classified as oxidation-reduction? What, for example, is a reducing flame; an oxidizing atmosphere?

2. How do you systematically balance equations for redox reactions? How do you recognize what is oxidized, what is reduced?

3. What occurs in an electrolytic cell; in a voltaic cell? How do you determine what voltage must be applied or what can be obtained?

4. What factors determine the voltage of a given cell reaction? How, for example, does the voltage depend on concentration, T and P? Again, why does the voltage drop during the use of a voltaic cell? Why do batteries run down?

5. What are some of the practical voltaic cells now in use? How do you decide what materials can be used for the electrodes in these cells?

6. What energy effects accompany redox reactions, whether or not they occur in electrochemical cells? What are some of the applications of redox reactions outside electrical power generation?

7. What can you say about the rates and extents of redox reactions? About how these depend on reaction conditions?

8. What correlations can you make between, say, voltage and chemical periodicity? Or between the ease of reducing a given species and its ionization energy or electronegativity?

YOU WILL NEED TO KNOW

Concepts

1. How to write the Lewis structure for a molecule or ion. See Skill 2 in Chapter 10 of this guide.

2. The general properties of acidic and basic water solutions. See Section 17.1 in the text.

Math

1. How to write and balance chemical equations; how to perform stoichiometric calculations, particularly involving concentrations (M). See Chapter 4.

2. How to determine logarithms and antilogs; how to work with pH. See Appendix 4 and Skill 2 in Chapter 17.

BASIC SKILLS

1. Given the formula of a molecule or ion, determine the oxidation number of each atom.

The rules for assigning oxidation numbers are given in the text and applied in Example 21.1. See the catalog of problems for further illustration.

You should keep in mind that oxidation numbers are assigned in a quite arbitrary manner and do not have any direct physical meaning. Thus, you should not be disturbed if you find that an atom in a molecule or polyatomic ion has an oxidation number of 0 or $-\frac{1}{2}$.

2. Given the formulas of products and reactants, balance a redox equation by the half-equation method.

The method is described in the text in considerable detail. Perhaps another example would be helpful.

－－－－－－－－－－－－－－－－－－－－－－－－－－－－－－－－－－－

Balance the equation for the reaction

$$Sn^{2+}(aq) + NO_3^-(aq) \rightarrow Sn^{4+}(aq) + NO_2(g)$$

first in acidic and then in basic solution. _____ ; _____
The oxidation half-equation is:

$$Sn^{2+}(aq) \rightarrow Sn^{4+}(aq) + 2e^-$$

where the two electrons are required to balance the charges. The reduction half-equation involves the NO_3^- ion:

$$NO_3^-(aq) + 2 H^+(aq) + e^- \rightarrow NO_2(g) + H_2O$$

Here, it was necessary to add one H_2O molecule to balance oxygen. This in turn required the addition of 2 H^+ to the left to balance hydrogen. Finally, one electron was added to the left to balance charges.

To obtain the overall equation in acidic solution, we multiply the second half-equation by 2 and add it to the first. This has the effect of "canceling out" the electrons:

$$Sn^{2+}(aq) \rightarrow Sn^{4+}(aq) + 2\ e^-$$

$$2\ NO_3^-(aq) + 4\ H^+(aq) + 2\ e^- \rightarrow 2\ NO_2(g) + 2\ H_2O$$

$$Sn^{2+}(aq) + 4\ H^+(aq) + 2\ NO_3^-(aq) \rightarrow Sn^{4+}(aq) + 2\ NO_2(g) + 2\ H_2O$$

To balance the equation in basic solution, we add 4 OH^- ions to both sides:

$$Sn^{2+}(aq) + 2\ NO_3^-(aq) + 4\ H_2O \rightarrow Sn^{4+}(aq) + 2\ NO_2(g) + 2\ H_2O + 4\ OH^-(aq)$$

which simplifies to:

$$Sn^{2+}(aq) + 2\ NO_3^-(aq) + 2\ H_2O \rightarrow Sn^{4+}(aq) + 2\ NO_2(g) + 4\ OH^-(aq)$$

- -

A couple hints:
— O_2 in water solution is ordinarily reduced to either H_2O (acidic solution) or OH^- (basic solution).
— It is quite possible for the same species to be both oxidized and reduced (that is, to undergo disproportionation). See Section 22.2 in the text.

3. **Given a balanced redox equation, label the oxidizing and reducing agents; the species being oxidized and reduced.**

Example 21.2 illustrates how this is done by following the change in oxidation numbers. In particular, a species undergoing oxidation shows an increase in oxidation number. That same species acts as a reducing agent. A species undergoing reduction shows a decrease in oxidation number while acting as an oxidizing agent.

4. **Given the redox reaction occurring in an electrochemical cell:**
   a. **label the anode and cathode;**
   b. **write the half-equation for the reaction occurring at each electrode;**
   c. **describe the electron and ion flow;**
   d. **give the cell notation.**

Example 21.3 does this (parts a–c) for the reaction occurring in a voltaic cell. The exercise following the example gives the cell notation. Electrolytic cells have already been described in some detail, for example in Chapter 5.

Note that in any cell, whether voltaic or electrolytic, **oxidation occurs at the anode, reduction at the cathode.** Within a cell, anions move towards the anode, cations toward the cathode. Outside the cell, electrons move out of the anode and into the cathode.

In designing a cell it is frequently necessary to use an inert, unreactive electrode. Platinum is a safe (though expensive!) choice. See the catalog of problems for additional practice in this skill.

5. **Use standard voltages (Table 21.1) to:**
   a. **compare the strengths of different oxidizing agents; different reducing agents.**

Referring to Table 21.1, the species in the left column can all, at least in principle, act as oxidizing agents. As you move down the column, from $Li^+(aq)$ at the top to $F_2(g)$ at the bottom, E^0_{red} becomes more positive (from -3.05 V for Li^+ to $+2.87$ V for F_2) and so oxidizing strength increases. That is, the species become more easily reduced.

Species in the right-hand column of Table 21.1 are all possible reducing agents. As you move up this column, from $F^-(aq)$ at the bottom to $Li(s)$ at the top, the E^0_{ox} becomes more positive (from -2.87 V for F^- to $+3.05$ V for Li) and so reducing strength increases. That is, the species become more easily oxidized.

See Example 21.4 for an application of this skill.

b. **calculate a cell voltage at standard concentration and pressure at 25°C.**

The relationship here is simple. The standard cell voltage, E^0, is $E^0 = E^0_{red}$ for the species reduced $+ E^0_{ox}$ for the species oxidized. It can be used to calculate the maximum voltage which is produced by the spontaneous reaction going on in a voltaic cell (see Example 21.5). It can also be used to calculate the minimum voltage which must be applied to carry out a nonspontaneous reaction in an electrolytic cell. See the discussion directly following Example 21.5 for an illustration of this point.

c. **decide whether or not a given redox reaction will occur at standard concentration and pressure at 25°C.**

The principle here is simple. If the calculated E^0 is positive, the reaction is spontaneous. It would, for example, occur if the species were mixed in a beaker or test tube; alternatively, it could serve as a source of energy in a voltaic cell. If the calculated E^0 is negative, the reaction is nonspontaneous. Work would have to be done to make the reaction go. It could, for example, be made to take place by supplying electrical energy in an electrolytic cell.

Example 21.6 illustrates this skill. Three points which you should keep in mind in working problems of this type are:

— You must combine an oxidation half-reaction with a reduction half-reaction. Students sometimes attempt to combine two oxidations or two reductions, thereby obtaining an absurd answer.

— Sometimes there will be more than one combination that will give a positive E^0. When this happens, there will be competing spontaneous reactions; the one which occurs most rapidly (not necessarily the one with the most positive E^0 value) will predominate.

— You must use as reactants only those species which are present at relatively high concentrations. For example, in 1 M strong acid solution, you cannot expect OH^- to be involved as a reactant in either oxidation (to O_2) or reduction (to H_2). On the other hand, a species such as H^+ might very well undergo reduction in such a solution.

6. **For a given redox reaction, write the expression for the Nernst equation and use the equation to calculate:**

a. **the voltage E of a cell, given E^0 and the concentrations of all species.**

The general expression for the Nernst equation is presented in the text (Equation 21.16). Note that concentration terms for solids and liquids do not appear in the expression. Note too that n represents the total number of moles of electrons transferred for the balanced redox equation as written.

Example 21.7 and the exercise that follows it call for the application of this skill.

- -

What is the value of n for the redox equation (unbalanced):

$$Sn^{2+}(aq) + H^+(aq) + NO_3^-(aq) \rightarrow Sn^{4+}(aq) + NO_2(g) + H_2O _____\ ?$$

This equation was balanced in Skill 2, above. Suppose you start the process of balancing again. First, you assign oxidation numbers in order to see what species undergoes reduction, what species undergoes oxidation. Tin is obviously oxidized. Nitrogen is reduced (from +5 to +4). So what is the value of n? The final balanced equation given above is for one mole of tin reacting with two moles of nitrate. The total number of moles of electrons transferred is two: n = 2.

- -

b. **the concentration of one species, given E, E^0, and the concentrations of all other species.**

This is a slight twist on the above skill. See the catalog of problems for an illustration of this skill

SELF-TEST

True or False

1. In a redox reaction, the oxidizing agent gains electrons.   ( )

2. A redox equation balanced for acidic solution can be ( )
converted to apply to basic solution by adding the proper number of
OH^- ions to both sides.

3. In the electrolysis of a solution of $Ag(S_2O_3)_2{}^{3-}$, three ( )
moles of electrons are required to form a mole of Ag.

4. In any cell, the cathode is the electrode where reduction ( )
occurs.

5. The strongest oxidizing agents have the largest, most ( )
positive E^0_{red}.

6. Since E^0_{red} is larger, more positive, for F_2 than for Ag^+, ( )
F^- must be a better reducing agent than Ag.

7. The voltage of a cell in which the reaction occurring is   ( )

$$2\ Ag^+(aq) + Cu(s) \rightarrow 2\ Ag(s) + Cu^{2+}(aq)$$

at standard conditions will be: $E^0_{ox}(Cu) + 2\ E^0_{red}(Ag^+)$.

Multiple Choice

8. The oxidation number of Mn in the MnO_4^- ion is   ( )
    (a) -2 (b) $+6$
    (c) $+7$ (d) $+8$

9. The oxidation number of P in H_3PO_4 is   ( )
    (a) -3 (b) $+1$
    (c) $+3$ (d) $+5$

10. The oxidation number of nitrogen in hydrazoic acid, HN_3, ( )
is
    (a) -3 (b) -1
    (c) $-1/3$ (d) $+1$

11. When the half-equation $I_2 + e^- \rightarrow I^-$ is balanced, the ( )
coefficients of I_2 e^-, and I^- are, respectively,
    (a) 1, 1, 1   (b) 1, 1, 2
    (c) 1, 2, 2   (d) 1, 0, 2

12. When the half-equation $HSO_3^- + H_2O \rightarrow SO_4^{2-} + H^+ + e^-$ is ( )
balanced, the coefficients, reading from left to right, are
    (a) 1, 1, 1, 1, 1   (b) 1, 1, 1, 2, 2
    (c) 1, 2, 1, 5, 2   (d) 1, 1, 1, 3, 2

13. When the half-equations in Questions 11 and 12 are ( )
combined, it is necessary to multiply the reduction half-equation by
_____ and the oxidation half-equation by _____ before adding.
    (a) 1, 1   (b) 1, 2
    (c) 2, 1   (d) 1, 3

14. Given the cell notation, $Ni | Ni^{2+} || (Pt) Sn^{4+} | Sn^{2+}$,   ( )
    (a) Ni is oxidized to Ni^{2+}(aq) at a nickel anode
    (b) Sn^{4+}(aq) is reduced to Sn^{2+}(aq) at an inert cathode
    (c) the overall reaction is $Ni(s) + Sn^{4+}(aq) \rightarrow Ni^{2+}(aq) +$
         Sn^{2+}(aq)
    (d) all the above

15. When a lead storage battery is charged, lead sulfate is   ( )
    (a) formed at the cathode
    (b) formed at the anode
    (c) formed at both electrodes
    (d) removed from both electrodes

16. The standard reduction voltages of Cl_2 and Cu^{2+} are ( )
+1.36 V and +0.34 V, respectively. The E^0 value for the reaction
$Cu^{2+}(aq) + 2 Cl^-(aq) \rightarrow Cu(s) + Cl_2(g)$ is
    (a) -2.38 V   (b) -1.70 V
    (c) -1.02 V   (d) $+1.70$ V

17. Which one of the following metals reacts with HNO_3 but ( )
not with dilute HCl?
    (a) Pt   (b) Mg
    (c) Na   (d) Cu

18. Which one of the following metals reacts with dilute HCl ( )
but not with water?
    (a) Ag   (b) Na
    (c) Ni   (d) Ca

19. It is possible to increase the voltage of a cell in which the ( )
reaction is $Zn(s) + Cu^{2+}(aq) \rightarrow Zn^{2+}(aq) + Cu(s)$ by increasing the
    (a) concentration of Zn^{2+}
    (b) concentration of Cu^{2+}

   (c)   size of the Zn electrode
   (d)   size of the Cu electrode

  20. In general, the voltage of a voltaic cell depends on the ( )
composition of the reactants and products, and on

(a) concentration	(b) pH
(c) temperature	(d) it may depend on all of these

Problems

  1. Consider the redox reaction: $4\ Cr^{2+}(aq) + O_2(g) + 4\ H^+(aq) \rightarrow 4\ Cr^{3+}(aq) + 2\ H_2O$.

   (a)   Assign oxidation numbers to all atoms.
   (b)   Indicate the species acting as an oxidizing agent.
   (c)   Indicate the species undergoing oxidation.

  2. Write balanced equations for the redox reactions:

   (a) $Co^{2+}(aq) + H^+(aq) + H_2O_2(aq) \rightarrow Co^{3+}(aq) + H_2O$
                                             (acidic solution)
   (b) $ClO_3^-(aq) + NO_2^-(aq) \rightarrow Cl^-(aq) + NO_3^-(aq)$ (basic solution)

  3. Suppose the spontaneous redox reaction occuring in a voltaic cell is $Sn(s) + Cl_2(g) \rightarrow Sn^{2+}(aq) + 2\ Cl^-(aq)$, $E^0 = +1.50$ V.

   (a)   Give the half-equation for the reaction occurring at each electrode.
   (b)   Describe the flow of ions inside the cell and electrons outside the cell.
   (c)   Give a reasonable notation for the cell.

  4. Consider again the reaction given for Problem 1, above. $E^0 = +1.64$ V for the reaction at 25°C.

   (a)   Write the expression for the Nernst equation.
   (b)   Determine the cell voltage if the concentration of H^+ is decreased to 1.0×10^{-7} M, while all other concentrations and pressures remain at 1. Is the reaction spontaneous under these conditions?

  5. Derive an expression that relates the cell voltage to the pH in the cell, operating at 25°C and 1 atm, and 1 M Cl^-:

$$H_2(g) + Cl_2(g) \rightarrow 2\ H^+(aq) + 2\ Cl^-(aq),\ E^0 = +1.36\ V.$$

SELF-TEST ANSWERS

  1. **T** (The oxidizing agent brings about oxidation of another species by removing electrons from that species.)

2. **T**

3. **F** (The oxidation number of Ag in the complex ion is +1. See Skill 1 of Chapter 19.)

4. **T** (And oxidation occurs at the anode in any cell.)

5. **T** (The strongest oxidizing agent is the most easily reduced.)

6. **F** (The reverse is true. Ag will have the larger, more positive E_{ox}^0.)

7. **F** (Do not multiply by 2. Rather, $E^0 = E_{ox}^0 + E_{red}^0$.)

8. **c** (Start by assigning oxygen its usual -2. See the rules in Section 21.1 of the text.)

9. **d** (Start with O; then H, with +1.)

10. **c** (Since H must be +1 and the sum over four atoms must be zero.)

11. **c** (First balance mass, the numbers of atoms, of the species being reduced; then balance charge. This one is easily done by inspection.)

12. **d** (First balance S; then O, using H_2O; H, using H^+; then charge, adding e^-.)

13. **a** (For electrical neutrality, electron loss must equal electron gain.)

14. **d** (Reading left to right, anode at the left and cathode at the right.)

15. **d** (While in use, i.e. discharging, $PbSO_4$ forms at both electrodes.)

16. **c** ($E^0 = E_{red}^0 Cu^{2+} + E_{ox}^0 Cl^- = E_{red}^0 Cu^{2+} - E_{red}^0 Cl_2$.)

17. **d** (Use Table 21.1 to determine E^0 for the possible redox reactions. NO_3^-, rather than H^+, is the oxidizing agent.)

18. **c** (H^+, not H_2O, is reduced by Ni. See Example 21.6 for this and the preceding question.)

19. **b** (Show that E increases by noting the effect of substituting a larger value of (conc. Cu^{2+}) in the Nernst equation.)

20. **d** (Many different redox reactions involve H^+ or OH^- as a species in the equation for the reaction — whether or not such species actually lose or gain electrons. The presence of H^+ or OH^- will be reflected in the Nernst equation expression and so determine E. Also note that T is $25°$ for all the calculations involving Table 21.1.)

Solutions to Problems

1. (a) Cr $(+2 \rightarrow +3)$; O $(0 \rightarrow -2)$; H $(+1)$
   (b) O_2 is the oxidizing agent. It is reduced (oxidation number decreases).
   (c) Cr^{2+} is oxidized.

2. (a) oxidation: $Co^{2+}(aq) \rightarrow Co^{3+}(aq) + e^-$
   reduction: $H_2O_2(aq) + 2 H^+(aq) + 2 e^- \rightarrow 2 H_2O$

   (O goes from -1 to -2)

overall: $2 Co^{2+}(aq) + 2 H^+(aq) + H_2O_2(aq) \rightarrow 2 Co^{3+}(aq) + 2 H_2O$

(b)   oxidation: $NO_2^-(aq) + 2 OH^-(aq) \rightarrow NO_3^-(aq) + H_2O + 2 e^-$
      (N goes from +3 to +5)
reduction: $ClO_3^-(aq) + 3 H_2O + 6 e^- \rightarrow Cl^-(aq) + 6 OH^-(aq)$
      (Cl goes from +5 to −1)

overall: $3 NO_2^-(aq) + ClO_3^-(aq) \rightarrow 3 NO_3^-(aq) + Cl^-(aq)$

3. (a)   anode (oxidation): $Sn(s) \rightarrow Sn^{2+}(aq) + 2 e^-$
      cathode (reduction): $Cl_2(g) + 2 e^- \rightarrow 2 Cl^-(aq)$

(b)   Anions (Cl^-) move toward the anode, away from the cathode where they are formed. Cations (Sn^{2+}) move toward the cathode, away from the anode where they are formed. Electrons flow out of the anode and into the cathode.

(c) $Sn \mid Sn^{2+} \parallel (Pt)Cl_2 \mid Cl^-$

4. (a) $E = E^0 - \dfrac{0.0591}{n} \text{Log} \dfrac{(\text{conc. } Cr^{3+})^4}{(\text{conc. } Cr^{2+})^4 P_{O_2} (\text{conc. } H^+)^4}$

Here, 4 mol Cr^{2+} lose 4 mol e^-. So, n = 4.

$E = 1.64 - \dfrac{0.0591}{4} \text{Log} \dfrac{(\text{conc. } Cr^{3+})^4}{(\text{conc. } Cr^{2+})^4 P_{O_2} (\text{conc. } H^+)^4}$

(b) $E = 1.64 - \dfrac{0.0591}{4} \text{Log} \dfrac{1}{(1.0 \times 10^{-7})^4}$

$= 1.64 - \dfrac{0.0591}{4} \text{Log } 10^{28}$

$= 1.64 - \dfrac{0.0591}{4} (28)$

$= +1.23 \text{ V}$

Since $E > 0$, the reaction is spontaneous.

5. $E = E^0 - \dfrac{0.0591}{2} \text{Log} \dfrac{(\text{conc. } H^+)^2 (\text{conc. } Cl^-)^2}{P_{H_2} P_{Cl_2}}$

$= 1.36 - \dfrac{0.0591}{2} \text{Log (conc. } H^+)^2$

$= 1.36 - \dfrac{0.0591}{2} [2 \text{ Log (conc. } H^+)]$

$= 1.36 + 0.0591 [- \text{Log (conc. } H^+)]$

$= 1.36 + 0.0591(\text{pH})$

SELECTED READINGS

An interesting variation on a common oxidation-reduction reaction:

Walker, J., Flames in Which Air Is Introduced into a Flammable Gas Rather than Vice Versa, *Scientific American* (November 1979), pp. 192–200.

Selected voltaic cells and electrical power plants:

Fickett, A.P., Fuel-Cell Power Plants, *Scientific American* (December 1978), pp. 70–76.

Johnston, W.D., The Prospects for Photovoltaic Conversion, *American Scientist* (November-December 1977), pp. 729–736.

Lawrence, R.M., Electrochemical Cells for Space Power, *Journal of Chemical Education* (June 1971), pp. 359–361.

Weissman, E.Y., Batteries: The Workhorses of Chemical Energy Conversion, *Chemistry* (November 1972), pp. 6–11.

Wrighton, M.S., Photochemistry, *Chemical and Engineering News* (September 3, 1979), pp. 29–47.

22

REDOX REACTIONS; OXIDATION
NUMBERS OF THE ELEMENTS

QUESTIONS TO GUIDE YOUR STUDY

1. How do you recognize what is oxidized and what is reduced in the equation for a redox reaction?

2. What are the oxidation states (numbers) of the elements as they commonly occur in nature, or are encountered in the laboratory?

3. How do metals differ from nonmetals? Are there trends in oxidation numbers within a group or period in the Periodic Table? Can a given element have more than one oxidation number? Are some states more stable than others?

4. For an atom of a given element, what is the maximum value for its oxidation number; the minimum value? How are these values related to electronic structure for a monatomic species?

5. What reaction conditions favor one oxidation state and not another?

6. Can you relate non-redox properties, such as the relative strengths of the acids $HClO_4$ and $HClO$, to oxidation numbers?

7. Can the Nernst equation be applied to a half-reaction as well as to an overall reaction? How do reaction conditions, particularly concentration, affect the voltage assigned to a half-reaction?

8. What is the nomenclature that spells out the oxidation state of an element in any given compound or ion?

9. What is the mechanism by which the common redox reaction of corrosion occurs? How is the reaction speeded up, slowed down, or prevented?

10. What are some of the common oxidizing agents, in the home as well as in the laboratory? What are their uses? How do they work (what are the reactions, what are the products)?

YOU WILL NEED TO KNOW

See this section of Chapter 21 in this guide.

This chapter is largely an extension of the preceding one. You will need to know essentially all the concepts and math required for Chapter 21 as well as the principles developed in that chapter. In addition:

Concepts

1. The general nature of the Periodic Table — the location of the metals, nonmetals, etc.

Math

1. The nature of the dissociation equilibria of weak acids and bases. How they are affected by a change in pH. The nature of step-wise dissociation of weak acids. See Chapter 18.

2. The nature of a titration. See Chapters 4 and 17.

BASIC SKILLS

Most of the skills used in this chapter were discussed in Chapter 21 of the guide. The following illustrate their applications here:

— *Write a balanced equation for a redox reaction.* Examples 22.2 and 22.4 review this skill. Note that part of this process, writing a balanced half-equation, is required to work Example 22.6 as well.

The reaction considered in Example 22.4, a **disproportionation**, is new to this chapter. In such a reaction, some atoms of a given element are oxidized while other atoms of the element are reduced. Both sets of atoms originate in one and the same reactant species. Disproportionation is discussed in detail in the body of the text and serves to illustrate two of the worked examples in this chapter (22.1 and 22.4). You should realize that this kind of redox reaction is likely to occur only when an element is in an intermediate oxidation state.

— *Use standard voltages (Table 21.1 and Figures 22.8, 22.9, and 22.11) to calculate a cell voltage under standard conditions.* Decide whether or not the reaction is spontaneous. See Skill 5 in Chapter 21. Examples 22.1, 22.4, and 22.5 apply this skill here.

— *For a given redox reaction, write the expression for the Nernst equation and use it to calculate the voltage E of a cell, given E^0 and the concentrations of all species.* Example 22.5 applies this skill (6a in the preceding chapter).

Skills new to this chapter include the following:

1. For a given half-reaction, write the expression for the Nernst equation. Use it to calculate the half-cell voltage, given the standard voltage and the concentrations of all species.

No new principle is involved here. This application of the Nernst equation is entirely analogous to the application mentioned just above. See Examples 22.3 and 22.6.

2. Given the balanced equation for a redox reaction and titration data for the reaction, calculate the concentration of one of the reactant species.

See Example 22.2. Again, no new principle is required here. The approach is the same as that used with acid-base reactions (Chapters 4 and 17).

3. Describe some of the more common oxidation states and reactions of O, Cl, N, and S; the oxyanions of these elements and of Cr and Mn. Describe the reactions and mechanism of corrosion.

There is a considerable amount of descriptive material introduced in this chapter. You should be familiar with some of these reactions and how they are affected by changes in concentration, especially pH.

SELF-TEST

True or False

1. For an atom of a nonmetallic element, the lowest oxidation   (   )
number is expected to be the charge on a monatomic anion.

2. For a metal of Group 2 in the Periodic Table, the highest   (   )
oxidation number is expected to be +2.

3. Variable oxidation states are characteristic of only the   (   )
transition metals.

4. You would expect that H_2SO_4, unlike HNO_3, could not act   (   )
as an oxidizing agent.

5. Negative oxidation numbers are characteristic of only the   (   )
nonmetals.

6. The oxidizing strength of an oxyanion is ordinarily greatest   (   )
at low pH.

Multiple Choice

7. Oxygen is *not* in its common oxidation state of -2 in   ( )
   (a) OF_2 (b) H_2O_2
   (c) KO_2 (d) any of these

8. In which of the following species does Cl have an oxidation   ( )
number of +5?
   (a) Cl^- (b) ClO^-
   (c) ClO_3^- (d) ClO_4^-

9. Which of the species below would be called chlorite?   ( )
   (a) ClO^- (b) ClO_2^-
   (c) ClO_3^- (d) ClO_4^-

10. In the series of oxyacids: $HOCl$, $HOClO$, $HOClO_2$, and   ( )
$HOClO_3$, the strongest acid is the one where Cl has the oxidation
number
   (a) +1   (b) +3
   (c) +5   (d) +7

11. Which of the following species can function only as an   ( )
oxidizing agent, and not as a reducing agent?
   (a) NO_3^- (b) NO_2^-
   (c) NH_3 (d) all of these

12. Disproportionation is unlikely for the species   ( )
   (a) NO_3^- (b) NO_2^-
   (c) N_2 (d) all of these

13. Conversion of CrO_4^{2-} to $Cr_2O_7^{2-}$ is favored by   ( )
   (a) low pH   (b) high pH
   (c) strong oxidizing agents   (d) strong reducing agents

14. To take advantage of the oxidizing strength of $Cr_2O_7^{2-}$, ( )
you should use it in _____ solution.
   (a) acidic   (b) basic
   (c) neutral   (d) dilute

15. A solution of NaOH saturated with Cl_2 at room tempera- ( )
ture will contain appreciable concentrations of all but one of the
following species. Indicate the exception.
   (a) Cl^- (b) OH^-
   (c) ClO^- (d) ClO_4^-

16. Corrosion ordinarily occurs more readily in seawater than in ( )
fresh water because
   (a) Na^+ ions in seawater attack iron
   (b) Cl^- ions in seawater attack iron

(c)   seawater is a better electrical conductor

(d) O_2 is more soluble in seawater

17. In order of increasing ease of oxidation, the halides are   (   )
   (a) $Cl^- < Br^- < I^-$ (b) $I^- < Br^- < Cl^-$
   (c)   all are easily oxidized   (d)   none can be oxidized

18. Since perchloric acid is potentially explosive, you would   (   )
never add concentrated H_2SO_4 to anything suspected of containing
   (a) ClO_4^- (b) ClO^-
   (c) Cl^- (d)   any of these

19. Consider the half-equation balanced for acidic solution: (   )

$$HNO_2(aq) + 7\ H^+(aq) + 6\ e^- \rightarrow NH_4^+(aq) + 2\ H_2O$$

If this were balanced for basic solution, you would expect to use
   (a) NO_2^- instead of HNO_2 (b) NH_3 instead of NH_4^+
   (c) OH^- instead of H^+ (d)   all the above

20. Given the half-reaction: (   )

$$H_2O_2(aq) + 2\ H^+(aq) + 2\ e^- \rightarrow 2\ H_2O,\ E^0_{red} = +1.77\ V$$

you would expect H_2O_2 to be most extensively reduced at
   (a)   low pH   (b)   high PH
   (c)   pH = 7   (d)   the anode

Problems

1. Write balanced equations for:
   (a)   the disproportionation of phosphorus, P_4, to PH_3 and HPO_3^{2-}
           in basic solution.
   (b)   the "dissolving" of chlorine in aqueous NaOH to give chloride
           and hypochlorite.

2. Consider the redox reaction:

$$10\ Cl^-(aq) + 2\ MnO_4^-(aq) + 16\ H^+(aq) \rightarrow 5\ Cl_2(g) + 2\ Mn^{2+}(aq) + 8\ H_2O$$

   (a)   What is the cell voltage under standard conditions?
   (b)   Would the reaction be spontaneous at pH = 7 if all other
           concentrations and pressures were 1?
           Standard voltages at 25°C:
           $Cl_2(g) + 2\ e^- \rightarrow 2\ Cl^-(aq)$ E^0_{red} = +1.36 V
           $MnO_4^-(aq) + 8\ H^+(aq) + 5\ e^- \rightarrow Mn^{2+}(aq) + 4\ H_2O$ E^0_{red} = +1.52 V

3. Give the formula of the species stable in acidic and basic solutions for:

	Acidic	Basic
(a) $+7$ Cl	_____	_____
(b) -2 S	_____	_____
(c) $+3$ N	_____	_____

4. Given the following standard reduction voltages (acidic solution):

$$\text{ClO}_4^- \xrightarrow{+1.19} \text{ClO}_3^- \xrightarrow{+1.21} \text{HClO}_2 \xrightarrow{+1.65} \text{HClO} \xrightarrow{+1.63} \text{Cl}_2 \xrightarrow{+1.36} \text{Cl}^-$$

(with $+1.47$ spanning ClO_4^- to HClO_2; $+1.43$ spanning ClO_3^- to HClO; $+1.49$ spanning HClO to Cl^-)

show by calculation whether each of the following disproportionation reactions is spontaneous (standard concentrations).

(a) $2\ \text{HClO}_2(aq) \rightarrow \text{ClO}_3^-(aq) + \text{HClO}(aq) + \text{H}^+(aq)$
(b) $3\ \text{HClO}(aq) \rightarrow \text{HClO}_2(aq) + \text{Cl}_2(g) + \text{H}_2\text{O}$
(c) $\text{Cl}_2(g) + 3\ \text{H}_2\text{O} \rightarrow \text{ClO}_3^-(aq) + 5\ \text{Cl}^-(aq) + 6\ \text{H}^+(aq)$

5. Suppose a certain redox reaction involves the reduction of O_2, at a constant pressure of 1 atm, and H^+, initially at 1 M, to water at $25°\text{C}$:

$$\text{O}_2(g) + 4\ \text{H}^+(aq) + 4\ e^- \rightarrow 2\ \text{H}_2\text{O},\ E^0 = +1.23\ \text{V}$$

As reaction proceeds, H^+ is consumed and the half-cell voltage for this reduction becomes smaller.

(a) At what concentration of H^+ has the E_{red} dropped to 0.40 V?
(b) Rewrite the half-equation so that it is more appropriate to these conditions (where the half-cell voltage is 0.40 V).

SELF-TEST ANSWERS

1. **T** (This would correspond to the charge of an anion with a noble gas structure. Examples: O^{2-} and Cl^-.)
2. **T** (As for the cations of charge +2, with noble gas structures.)
3. **F** (Most nonmetals also have more than one oxidation state. See Figure 22.1.)
4. **F** (The sulfur can certainly be reduced, for example to SO_3^{2-} or SO_2. Of course, H^+ could also act as an oxidizing agent. See Figure 22.11.)
5. **T** (As expected for the more electronegative elements.)

6. T   (High concentration of H^+ favors this reduction. See the discussion of the many oxyanions in this chapter of the text.)

7. d   (The oxidation number of O is +2 in OF_2; -1 in H_2O_2; $-\frac{1}{2}$ in KO_2.)

8. c   (Assign O an oxidation number of -2. See the rules in Section 21.1 of the text.)

9. b   (The others are hypochlorite, chlorate, and perchlorate. See Section 22.4.)

10. d   (That is, in $HOClO_3$ (or $HClO_4$), which you should recognize as one of the common strong acids. See Chapter 17.)

11. a   (In NO_3^-, N is in its highest oxidation state and so cannot be oxidized.)

12. a   (NO_3^- cannot be oxidized, as noted above. An intermediate oxidation state is required for disproportionation.)

13. a   (Can you write the balanced equation for this non-redox reaction?)

14. a   (See the preceding question. Note that $Cr_2O_7{}^{2-}$ is a stronger oxidizing agent than $CrO_4{}^{2-}$.)

15. d   (A result of disproportionation. See Problem 1b below.)

16. c

17. a   (As confirmed by their relative positions in Table 21.1.)

18. a   ($HClO_4$ is perchloric acid. Nor would you add concentrated H_2SO_4 to a chlorate!)

19. d   (These are all basic species: NO_2^-, NH_3, and OH^-.)

20. a   (Raising the concentration of H^+ makes this reaction "more spontaneous" by increasing E. Again, an application of the Nernst equation.)

Solutions to Problems

1. (a) $P_4(s) + 12\ H^+ + 12\ e^- \rightarrow 4\ PH_3(g)$

$$\underline{P_4(s) + 12\ H_2O \rightarrow 4\ HPO_3{}^{2-}(aq) + 20\ H^+(aq) + 12\ e^-}$$

$$2\ P_4(s) + 12\ H_2O \rightarrow 4\ PH_3(g) + 4\ HPO_3{}^{2-}(aq) + 8\ H^+(aq)$$

or, $P_4(s) + 6\ H_2O \rightarrow 2\ PH_3(g) + 2\ HPO_3{}^{2-}(aq) + 4\ H^+(aq)$
$$\underline{\qquad\qquad +4\ OH^-(aq) \qquad\qquad\qquad\qquad\qquad +4\ OH^-(aq)}$$

$$P_4(s) + 2\ H_2O + 4\ OH^-(aq) \rightarrow 2\ PH_3(g) + 2\ HPO_3{}^{2-}(aq)$$

(b) $Cl_2(g) + 2\ OH^-(aq) \rightarrow Cl^-(aq) + ClO^-(aq) + H_2O$

(Equation 22.22)

2. (a) $E^0 = E_{red}^0(MnO_4^-) + E_{ox}^0(Cl^-)$
$= 1.52 - 1.36$
$= +0.16$ V

(b) The reaction is spontaneous if $E > 0$. At pH = 7, H^+ conc. = 10^{-7} M, so, using the Nernst equation:

$$E = E^0 - \frac{0.0591}{n} \text{Log}[1/(\text{conc. } H^+)^{16}]$$

Here n = 10.

$$E = 0.16 - \frac{0.0591}{10} \text{Log } [1/10^{-7})^{16}]$$

$$= 0.16 - \frac{0.0591}{10} \text{Log } 10^{112}$$

$$= 0.16 - \frac{0.0591}{10} (112)$$

$$= 0.16 - 0.66$$

$$= -0.50 \text{ V} \qquad \text{So the reaction is nonspontaneous.}$$

3. (a) ClO_4^- ClO_4^-
    (b) H_2S HS^-, S^{2-}
    (c) HNO_2 NO_2^-

4. (a) $E^0 = +1.65$ V $- 1.21$ V $= +0.44$ V; spontaneous
    (b) $E^0 = +1.63$ V $- 1.65$ V $= -0.02$ V; nonspontaneous
    (c) $E^0 = +1.36$ V $- 1.47$ V $= -0.11$ V; nonspontaneous

5. (a) $E = E^0 - \dfrac{0.0591}{4} \text{Log } (\text{conc } H^+)^{-4}$

    $0.40 = 1.23 + 0.0591 \text{ Log } (\text{conc. } H^+)$
    $\text{Log } (\text{conc. } H^+) = -14.0$
    $(\text{conc. } H^+) = 1 \times 10^{-14}$ M

    (b) Note that here (conc. OH^-) = 1 M. So, rather than H^+, you should show OH^- in the balanced equation. Adding 4 OH^- to both sides of the given equation gives:
    $O_2(g) + 2 H_2O + 4 e^- \rightarrow 4 OH^-(aq)$, $E^0 = 0.40$

SELECTED READINGS

See the Readings listed for the preceding chapter, and:

Fischer, R.B., Ion-Selective Electrodes, *Journal of Chemical Education* (June 1974), pp. 387–390.

Slabaugh, W.H., Corrosion, *Journal of Chemical Education* (April 1974), pp. 218–220.

Taube, H., Mechanisms of Oxidation-Reduction Reactions, *Journal of Chemical Education* (July 1968), pp. 452–461.

23

CHEMICAL THERMODYNAMICS; ΔH, ΔS, AND ΔG

QUESTIONS TO GUIDE YOUR STUDY

1. What do you mean by reaction spontaneity? What do you imply about the rate and extent of a reaction when you say that it is spontaneous?

2. Can you decide whether or not a particular reaction will occur without even trying to carry out the reaction? (You have been able to calculate ΔH, or K, or E for such reactions.)

3. What physical meaning do you associate with the quantities ΔH, ΔS, and ΔG for a reaction? What about the quantities of H, S, and G as they apply to individual substances? What kinds of measurements do you make to determine their values?

4. How are these thermodynamic quantities related to the masses of substances taking part in a reaction? Does Hess's Law apply to ΔG and ΔS as well as ΔH?

5. How are these thermodynamic quantities related to each other? How is each of them related to the properties of atoms and molecules, such as bond energy and molecular structure?

6. How do ΔH, ΔS, and ΔG depend on reaction conditions such as T, P, and concentration? Can you predict the direction and magnitude of the effect of a change in conditions for ΔH, ΔS, and ΔG?

7. Can you predict the sign or relative magnitude of ΔS for a given reaction? Likewise for either ΔH or ΔG?

8. What happens at the molecular level when the entropy of a system increases; when the enthalpy decreases; when the free energy decreases?

9. What is the thermodynamic criterion for equilibrium? (You have seen how the cell voltage may signal the end of a reaction.) How do you interpret equilibrium in terms of the "driving forces" behind changes in chemical systems?

10. To what kinds of systems, chemical or otherwise, can you apply the principles of spontaneity developed in this chapter?

YOU WILL NEED TO KNOW

Concepts

1. How to interpret the sign of ΔH, and ΔH itself, in terms of the making and breaking of bonds. Reviewed in the first section of the current chapter, but also see Chapter 6.

2. The nature of equilibrium phase changes. That is, recognize T_f and T_b as **equilibrium** temperatures at P = 1 atm. See Chapter 11.

Math

1. How to work problems in stoichiometry. See Chapter 4.

2. How to calculate ΔH for any given reaction. See Skill 3 in Chapter 6 of this guide.

3. How to calculate E^0 for any given redox reaction. See Skill 5 in Chapter 21 of the guide.

4. How to recognize and use the various equilibrium constants for reactions occurring in aqueous solution: K_{sp}, K_w, K_a, K_b, and K_d. See Chapters 17–19, and Skills 1 and 7 in Chapter 18 of the guide.

BASIC SKILLS

1. Given the standard molar entropies of reactants and products, calculate the standard entropy change, $ΔS^0$, for a reaction.

The calculation here is similar to the use of heats of formation in calculating ΔH for a reaction. (See Skill 3 in Chapter 6.) Note, however, that **all elements**, as well as compounds, **have finite positive values of S^0** at 25°C. The only zero value of the standard entropy is assigned to $H^+(aq)$. See Examples 23.1 and 23.2. See also the catalog of problems at the end of the chapter.

You should be able to predict and interpret the sign of the entropy change for many reactions. The general principle here is that in going from a more ordered state to a less ordered one, entropy increases. That is, $ΔS^0$ is positive. Examples of such processes include melting, vaporization, and the formation of a solution. For a chemical reaction, you can be sure that entropy increases if the number of moles of gas increases. For reactions not involving gases, ΔS is usually positive if there is a large increase in the number of moles.

2. **Given or having calculated ΔH and ΔS^0 for a reaction, calculate ΔG^0 for the reaction:**
    a.   **at 25°C.**
    b.   **at any temperature T.**

The first of these skills is illustrated by Example 23.2. The relationship between the three thermodynamic properties is given by the Gibbs-Helmholtz equation. In particular, at 1 atm and 1 M concentrations, this is Equation 23.12 in the text: $\Delta G^0 = \Delta H - T\Delta S^0$.

Example 23.3 illustrates the application of the second skill. See the problems for additional practice.

You should be able to interpret the sign of ΔG^0 for any reaction. In particular, a positive sign means the reaction is nonspontaneous. The reverse reaction would spontaneously occur. If the reaction is to go in the forward direction, work would have to be done.

A negative sign means the reaction is spontaneous. It will occur of its own accord. In fact, if set up in a suitable manner (perhaps in an electrochemical cell), it can be harnessed to do work for you. Note, however, nothing can be said about the rate of the reaction.

Finally, a ΔG^0 value of zero means that the reaction system is in a state of equilibrium.

Again, you should be able to predict the effect of a change in temperature on reaction spontaneity (that is, on the sign of ΔG^0), given the signs of ΔH and ΔS^0. This use of the Gibbs-Helmholtz equation is summarized by Table 23.5 in the text.

3. **Given ΔH and ΔS^0 for a reaction, calculate the temperature at which equilibrium will exist at 1 atm.**

As noted above, at the pressure of 1 atm equilibrium exists if $\Delta G^0 = 0$. Using the Gibbs-Helmholtz equation:

$$\Delta G^0 = 0 = \Delta H - T\Delta S^0 \text{; or, } T = \Delta H/\Delta S^0 \text{, and } \Delta S^0 = \Delta H/T$$

See Example 23.4 for an illustration of this skill. Also consider the following application.

- -

      Given the molar heat of vaporization of water at 100°C, $\Delta H_{vap} = 40.7$ kJ, calculate ΔS^0_{vap} at 100°C. _____
      At the boiling point, equilibrium exists and so $\Delta G^0 = 0$. So, using the relation derived above:

$$\Delta S^0 = \Delta H/T = 40.7/373 = 0.109 \text{ kJ/K or } 109 \text{ J/K.}$$

- -

4. Quantitatively relate ΔG^0 and E^0 for a given reaction at 25°C.

The relation here is $\Delta G^0 = -96.5 \, n \, E^0$, where the units of the free energy change is kilojoules, E^0 is expressed in volts, and n is the number of moles of electrons transferred in the reaction. See Example 23.5 for an application of this skill. Also see Skill 5b in Chaper 21.

5. Quantitatively relate ΔG^0 and K for a reaction in water solution.

At 25°C, the relation is $\Delta G^0 = -5.71 \, \text{Log}_{10} \, K$, where ΔG^0 is in kilojoules. Example 23.6 applies this skill. You will have to recognize the nature of the equilibrium constant you are dealing with. In the text example and the exercise that follows it, these are the constants K_{sp}, $1/K_w$, and K_a. Review Skills 1 and 7 in Chapter 18.

6. Apply the laws of thermochemistry to calculations involving ΔS^0 and ΔG^0, as well as ΔH.

A brief review of these laws as set out in Sections 6.2 and 6.3 in the text would be helpful here. The calculations involving ΔG and ΔS are entirely analogous to those you have done with ΔH.

— —

Given the reactions:

$$2 \, Sn(s) + O_2(g) \rightarrow 2 \, SnO(s), \qquad \Delta G^0 = -515 \, kJ$$

$$2 \, SnO(s) + O_2(g) \rightarrow 2 \, SnO_2(s), \qquad \Delta G^0 = -523 \, kJ$$

calculate ΔG^0 for $SnO_2(s) \rightarrow Sn(s) + O_2(g)$. _____

Suppose you start by adding the two given equations, and then cancel 2 SnO:

$$2 \, Sn(s) + 2 \, O_2(g) \rightarrow 2 \, SnO_2(s), \, \Delta G^0 = -515 + (-523) = -1038 \, kJ$$

Dividing through by 2, you now get:

$$Sn(s) + O_2(g) \rightarrow SnO_2(s), \text{ for which } \Delta G^0 = -1038/2 = -519 \, kJ$$

Finally, reversing the equation:

$$SnO_2(s) \rightarrow Sn(s) + O_2(g), \quad \Delta G^0 = +519 \, kJ$$

— —

SELF-TEST

True or False

    1. The standard molar entropy of an element at 25°C is zero. (   )

    2. Equilibrium can be considered as a balance struck between (   )
two opposing tendencies: A system tends to move toward a state of
minimum enthalpy; a system tends to move toward a state of
maximum entropy.

    3. For the decomposition of 1 mol $H_2O(l)$ to the elements at (   )
25°C, ΔG^0 = +237 kJ. This means that at least 237 kJ of work has to
be done to make this reaction go.

    4. If ΔG^0 for a reaction is positive, it is impossible to carry out (   )
the reaction unless either T or P is changed.

    5. All exothermic reactions become spontaneous as T ap- (   )
proaches 0 K.

    6. Only if ΔG^0 is close to zero will there be appreciable (   )
amounts of both reactants and products at equilibrium.

    7. For the reaction $PCl_5(g) \rightarrow PCl_3(g) + Cl_2(g)$, ΔS is positive. (   )

    8. If ΔH and ΔS are both negative for a given reaction, you (   )
expect ΔG to be negative at all temperatures.

Multiple Choice

    9. You would expect that for the formation of a compound (   )
from the elements at 25°C, 1 atm, ΔG^0
      (a) is usually positive
      (b) is usually negative
      (c) is usually close to zero
      (d) none of these is a reasonable generalization

    10. Which of the thermodynamic properties is essentially (   )
unaffected by a change in temperature?
      (a) ΔH (b) ΔS^0
      (c) ΔG^0 (d) both ΔH and ΔS^0

    11. For the reaction $Ag(s) + \frac{1}{2} Cl_2(g) \rightarrow AgCl(s)$, ΔG^0 = (   )
−110 kJ. For the reaction $2 AgCl(s) \rightarrow 2 Ag(s) + Cl_2(g)$, ΔG^0 is
      (a) −220 kJ   (b) −110 kJ
      (c) +110 kJ   (d) +220 kJ

12. Which one of the following statements best describes the ( )
relationship between ΔG^0 and temperature?
   (a) ΔG^0 is independent of T
   (b) ΔG^0 varies with T
   (c) ΔG^0 is a linear function of T
   (d) ΔG^0 usually decreases as T decreases

13. Which of the following would you expect to have the largest ( )
molar entropy at 25°C?
   (a) Li(g)   (b) Li(s)
   (c) Li(ℓ)   (d) they should be the same

14. For a certain reaction, ΔH is 85 kJ and ΔG^0 at 300 K is ( )
55 kJ. ΔS^0, in kJ/K, is about
   (a) +30   (b) +0.1
   (c) -0.1   (d) -30

15. Vaporization is an example of a process for which   ( )
   (a) ΔH, ΔS^0, and ΔG^0 are positive at all temperatures
   (b) ΔH and ΔS^0 are positive
   (c) ΔG^0 is negative at low T, positive at high T
   (d) ΔH is strongly dependent on the pressure

16. For a certain reaction, ΔH = + 2.5 kJ and ΔS^0 = ( )
0.010 kJ/K. This reaction will be at equilibrium at 1 atm at about
   (a) 0.25°C   (b) 25°C
   (c) -23°C   (d) cannot tell

17. Consider the reaction $C(s) + \frac{1}{2} O_2(g) \rightarrow CO(g)$, ΔH = ( )
-110 kJ. As the temperature is increased, ΔG^0 will
   (a) remain unchanged   (b) change sign
   (c) become less negative   (d) become more negative

18. The reaction $CaO(s) + H_2O(l) \rightarrow Ca(OH)_2(s)$ is spontaneous ( )
at 25°C; the reverse reaction becomes spontaneous at high temperature. This means that for the reaction listed above:
   (a) ΔH is +, ΔS^0 is +   (b) ΔH is +, ΔS^0 is -
   (c) ΔH is -, ΔS^0 is -   (d) ΔH is -, ΔS^0 is +

19. For a certain reaction, E^0 is positive. This means that   ( )
   (a) $\Delta G^0 > 0, K > 1$ (b) $\Delta G^0 > 0, K < 1$
   (c) $\Delta G^0 < 0, K > 1$ (d) $\Delta G^0 < 0, K < 1$

20. For a certain reaction, ΔG^0 is known at two different ( )
temperatures. From this information alone, you can calculate for the
reaction
   (a) ΔH (b) ΔS^0
   (c) both of these   (d) neither of these

Problems

1. Consider the reaction $CO(g) + \frac{1}{2} O_2(g) \rightarrow CO_2(g)$, $\Delta H = -283$ kJ.
Predict:
   (a) the sign of ΔS^0.
   (b) the effect of a change in temperature on reaction spontaneity.
   Is the reaction spontaneous at low or high T?

2. Using the molar heats of formation and standard entropies, at 25°C
and 1 atm:

	CuO(s)	$Cu_2O(s)$	$O_2(g)$
$\Delta H_f(kJ)$	-155	-167	0
$S^0(J/K)$	+44	+101	+205

calculate for the reaction $Cu_2O(s) + \frac{1}{2} O_2(g) \rightarrow 2\ CuO(s)$,
   (a) ΔH, ΔS^0, and ΔG^0 at 25°C. Interpret the signs.
   (b) the temperature at which equilibrium exists at 1 atm.

3. For the reaction $NH_3(aq) + H_2O \rightarrow NH_4^+(aq) + OH^-(aq)$ at 25°C,
$K_b = 1.77 \times 10^{-5}$. At 0°C, $K_b = 1.37 \times 10^{-5}$. Calculate ΔG^0 at 0°C and
25°C, using the relation: ΔG^0 (kilojoules) $= -0.0191T \log_{10} K$.

4. At 25°C, $E^0 = +0.16$ V for the redox reaction:

$$10\ Cl^-(aq) + 2\ MnO_4^-(aq) + 16\ H^+(aq) \rightarrow 5\ Cl_2(g) + 2\ Mn^{2+}(aq) + 8\ H_2O$$

   (a) Calculate ΔG^0 for this reaction at 25°C, using the relation:
   ΔG^0 (kilojoules) $= -96.5\ nE^0$.
   (b) Is this reaction spontaneous under standard conditions?

5. Use the data in Problem 3 above to determine ΔH and ΔS^0 for the
reaction given.

SELF-TEST ANSWERS

1. **F** (The only zero value at 25°C is an assigned one, that of
   $H^+(aq)$.)
2. **T** (As reflected by the combination of these two "driving forces":
   $\Delta G = \Delta H - T\Delta S$. The spontaneous movement of a system
   toward equilibrium is favored by a negative ΔH and a positive
   ΔS.)
3. **T** (If ΔG is positive, the reaction is nonspontaneous. The reverse
   reaction is spontaneous.)

4. **F** (See the preceding question. Work could be done to make it occur at T, 1 atm.)

5. **T** (From the Gibbs-Helmholtz relation, you see that ΔH determines the sign of ΔG at low T. That is, TΔS becomes less and less significant as T gets smaller.)

6. **T** (For such reactions, the system is already near a state of equilibrium with both reactants and products at 1 M, 1 atm. Also, K is close to 1. See Table 23.6 in the text.)

7. **T** (Primarily because Δn(gas) is positive.)

8. **F** (Use the Gibbs-Helmholtz relation to see that ΔG will be positive at high T. See Table 23.5 in the text.)

9. **b** (Reflecting the fact that most compounds are stable compared to the elements.)

10. **d** (This allows you to calculate ΔG as a function of T without worrying about any change in ΔH or ΔS.)

11. **d** (Like ΔH, both $ΔG^0$ and $ΔS^0$ are directly proportional to amount of reactants and products. Also, reversing a reaction changes the signs. See Skill 6.)

12. **c** (Since both ΔH and $ΔS^0$ change little with T, the Gibbs-Helmholtz equation has the form of a linear equation: $ΔG^0 = ΔH - TΔS^0 = b - aT$, where a and b are constants.)

13. **a** (Gaseous.)

14. **b** ($ΔS^0 = (ΔH - ΔG^0)/T = 30/300$.)

15. **b**

16. **c** ($ΔG^0 = 0 = ΔH - TΔS^0$; $T = ΔH/ΔS^0 = 2.5/0.010 = 250$ K.)

17. **d** (You should expect $ΔS^0$ to be positive since Δn(gas) > 0. Using the Gibbs-Helmholtz relation, both ΔH and $ΔS^0$ make $ΔG^0$ negative. $ΔS^0$ will make $ΔG^0$ still more negative at high T.)

18. **c** (You expect ΔH < 0 since it gives ΔG its sign at low T. ΔS determines the sign of ΔG at high T.)

19. **c** (A positive value of the cell voltage means the reaction is spontaneous. So ΔG < 0. For such reactions, more product forms, less reactant remains, at equilibrium. Having **started** at 1 M, 1 atm for all reactants and products, equilibrium must be described by K > 1.)

20. **c** (From two equations in two unknowns, each of the form of the Gibbs-Helmholtz equation, you can calculate the two unknowns. See Problem 5 below. Note that you could also then determine $ΔG^0$ at any T; E^0 and K as well!)

Solutions to Problems

1. (a) $ΔS^0$ should be negative, since Δn(gas) is negative.

(b) Both ΔH and ΔS^0 and negative. So, given $\Delta G^0 = \Delta H - T\Delta S^0$, you expect ΔG^0 to be negative at low T, positive at high T. (Reaction is spontaneous at low T, nonspontaneous at high T.)

2. (a) $\Delta H = \Sigma\Delta H_f$ (products) $- \Sigma\Delta H_f$ (reactants)

$= 2(-155) - (-167) = -143$ kJ

$\Delta S^0 = \Sigma S^0$ (products) $- \Sigma S^0$ (reactants)

$= 2(44) - [101 + \frac{1}{2}(205)] = -116$ J/K $= -0.116$ kJ/K

$\Delta G^0 = \Delta H - T\Delta S^0 = -143 - 298(-0.116) = -108$ kJ

From the sign of ΔH: product bonds are stronger than reactant bonds. ΔS^0 is negative, mostly because Δn(gas) is negative. The sign of ΔG^0 means the reaction is spontaneous at 25°C, 1 atm.

(b) $\Delta G^0 = 0 = \Delta H - T\Delta S^0$

$T = \Delta H/\Delta S^0 = -143/(-0.116) = 1230$ K or about 960°C

3. At 0° C, $\Delta G^0 = -0.0191(273)\log_{10} (1.37 \times 10^{-5}) = 25.4$ kJ

At 25°C, $\Delta G^0 = -0.0191(298)\log_{10} (1.77 \times 10^{-5}) = 27.1$ kJ

4. (a) $\Delta G^0 = -96.5$ n$E^0 = -96.5(10)(0.16) = -1.5 \times 10^2$ kJ

(b) The signs of both E^0 and ΔG^0 indicate the reaction is spontaneous under standard conditions.

5. $\Delta G^0 = \Delta H - T\Delta S^0$

At 25°C, $27.1 = \Delta H - 298\Delta S^0$

At 0°C, $25.4 = \Delta H - 273\Delta S^0$

Subtracting the second equation from the first gives:

$1.7 = -25\Delta S^0$, or, $\Delta S^0 = -0.068$ kJ/K

At 25°C: $\Delta G^0 = \Delta H - 298\Delta S^0$,

$\Delta H = \Delta G^0 + 298\Delta S^0 = 27.1 + 298(-0.068) = 6.8$ kJ

SELECTED READINGS

Alternative discussions of thermodynamics, from a highly organized outline of working principles to a detailed statistical approach which looks at molecular activity:

Bent, H.A., *The Second Law,* New York, Oxford University Press, 1965.

Campbell, J.A., *Why Do Chemical Reactions Occur?* Englewood Cliffs, N.J., Prentice-Hall, 1965.

MacWood, G.E., How Can You Tell Whether a Reaction Will Occur? *Journal of Chemical Education* (July 1961), pp. 334–337.

Mahan, B.H., *Elementary Chemical Thermodynamics,* New York, W.A. Benjamin, 1963.

Nash, L.K., *Chemthermo,* Reading, Mass., Addison-Wesley, 1972.

Pimentel, G.C., *Understanding Chemical Thermodynamics,* San Francisco, Holden-Day, 1969.

Porter, G., The Laws of Disorder, *Chemistry* (May 1968), pp. 23–25.

Sanderson, R.T., Principles of Chemical Reaction, *Journal of Chemical Education* (January 1964), pp. 13–22.

Other topics related to those of this chapter are found in:

Bent, H.A., Haste Makes Waste, Pollution and Entropy, *Chemistry* (October 1971), pp. 6–15.

Layzer, D., The Arrow of Time, *Scientific American* (December 1975), pp. 56–69.

Proctor, W.G., Negative Absolute Temperatures, *Scientific American* (August 1978), pp. 90–99.

Smith, W.L., Thermodynamics, Folk Culture, and Poetry, *Journal of Chemical Education* (February 1975), pp. 97–98.

24

NUCLEAR REACTIONS

QUESTIONS TO GUIDE YOUR STUDY

1. How are nuclear reactions different from "ordinary" chemical reactions? How would you experimentally recognize that a particular reaction was a nuclear reaction?

2. What factors determine whether a particular atom is radioactive? Are there correlations that can be made with the Periodic Table? With nuclear composition?

3. What are the properties of the various kinds of radiation? How, for example, do they interact with matter? What chemical reactions occur as a result of interaction with biological systems?

4. How do reaction conditions such as temperature, pressure, and concentration affect a nuclear reaction?

5. How would you experimentally determine the rate of a nuclear reaction? What can you say about the rate expression and reaction mechanism for a given nuclear reaction?

6. What can you say about the spontaneity and extent of a nuclear reaction?

7. What is the difference between "natural" and "artificial" nuclear reactions? Between fission and fusion?

8. What are some of the *chemical* applications of nuclear reactions?

9. What energy effects are associated with nuclear reactions? Can you predict whether a given nuclear reaction will be exothermic or endothermic?

10. What reactions occur in a nuclear power plant? What are some of the advantages and disadvantages of nuclear power? (How, for example, does the cost compare with that of conventional power from fossil fuels?)

YOU WILL NEED TO KNOW

Concepts

1. How to write and interpret nuclear symbols. See Skill 2 in Chapter 2 of the guide.

2. What is meant by *isotope* — Chapter 2.
3. The meaning of activation energy. See Chapter 14.

Math

1. How to work problems involving first order kinetics. See Skills 3 and 4 in Chapter 14 of this guide.

BASIC SKILLS

1. **Write a balanced equation for a nuclear reaction, given the identities of all but one of the reactants and products.**

The basic principle here is that both the total mass number (superscript at upper left) and the total nuclear charge (subscript at lower left) must "balance," i.e., must have the same value on both sides of the equation. This principle is illustrated in Examples 24.1 and 24.2. See the catalog of problems at the end of the chapter for additional practice in this skill.

2. **Use the first order rate equation and the expression for the half-life to relate the amount of a radioactive species to elapsed time.**

The pertinent equations here are those discussed in Chapter 14 in connection with first order reactions:

$$\log_{10} \frac{X_0}{X} = \frac{kt}{2.30} \; ; \; t_{\frac{1}{2}} = \frac{0.693}{k}$$

Examples 24.3 and 24.4 illustrate the application of these equations to radioactive processes. You should realize that no new concepts are introduced here. Very similar calculations were carried out in the problems and examples of Chapter 14. Review Skills 3 and 4 in that chapter of this guide.

3. **Given nuclear masses, calculate Δm for a nulcear reaction and relate it to the energy change, ΔE.**

Nuclear masses are given in the text in Table 24.4. Note that use of these masses leads directly to Δm in grams per mole of reactant. Thus, for the reaction $^{239}_{94}Pu \rightarrow ^{235}_{92}U + ^{4}_{2}He$, you would have, per mole of Pu reacting: Δm = 234.9934 + 4.0015 − 239.0006 or −0.0057 g. This difference in mass is readily converted to an energy difference by using Equation 24.16: $\Delta E(kJ) = -0.0057 \text{ g} \times 9.00 \times 10^{10} = -5.1 \times 10^{8}$ kJ.

Note that:

− This calculation gives Δm or ΔE per mole of reactant. If you want the change per gram of reactant, you must divide by the molar mass.

− A negative Δm and ΔE imply a spontaneous nuclear reaction. If Δm and ΔE are positive, the reaction is nonspontaneous. Such a reaction will not proceed of its own accord.

Calculations of this type are carried out in Examples 24.5 and 24.6.

4. Discuss the properties of the various kinds of radiation and their interactions with matter.

Besides knowing the symbols for the particles emitted in alpha, beta, gamma, and positron decay, you should know the masses and charges of these particles. Know too, for example, typical wavelengths for γ-rays. See Tables 24.2 and 24.3 for biological effects of radiation and typical exposure levels.

SELF-TEST

True or False

1. The emission of a β-particle leaves the atomic number ( ) unchanged but increases the mass number by one unit.

2. Emission of a positron is equivalent to the conversion of a ( ) proton to a neutron in the nucleus.

3. The most serious effect of long-term low-level radiation on ( ) the body is that it produces severe skin burns.

4. The longer the half-life of a radioactive isotope, the more ( ) rapidly it decays.

5. According to the Einstein relation, $\Delta E(kJ) = 9 \times$ ( ) $10^{10} \Delta m(g)$, the fusion of 1 g of deuterium would evolve 9×10^{10} kJ.

6. Probably the most important hazard associated with a ( ) nuclear power plant is the possibility of a *nuclear* explosion.

7. On the average, a person living in the United States is (   )
exposed to about twice as much radiation from natural sources as
from man-made sources.

Multiple Choice

8. The emission of an α-particle lowers the atomic number by (   )
_____ and the mass number by _____ , respectively.
    (a) 1, 1   (b) 1, 2
    (c) 2, 2   (d) 2, 4

9. Nuclear reactions differ from ordinary chemical reactions in (   )
all but one of the following ways. Indicate the exception.
    (a) The energy evolved per gram is much greater for
          nuclear reactions.
    (b) Nuclear reactions occur much more rapidly.
    (c) New elements are often formed in nuclear reactions.
    (d) In nuclear reactions, reactivity is essentially inde-
          pendent of the state of chemical combination.

10. Emission of which one of the following leaves both atomic (   )
number and mass number unchanged?
    (a) positron   (b) neutron
    (c) α-particle   (d) γ-radiation

11. A certain radioactive series starts with $^{235}_{92}$U and ends with (   )
$^{207}_{82}$Pb. In the overall process, _____ α-particles and _____ β-
particles are emitted.
    (a) 8, 6   (b) 14, 10
    (c) 7, 10   (d) 7, 4

12. Bombardment of $^{75}_{33}$As by a deuteron, $^{2}_{1}$H, forms a proton
and an isotope, which has a mass number of _____ and an atomic
number of _____ .
    (a) 73, 32   (b) 75, 33
    (c) 75, 32   (d) 76, 33

13. Which one of the following isotopes would be most likely to (   )
undergo fission?
    (a) $^{14}_{6}$C   (b) $^{59}_{27}$Co
    (c) $^{239}_{94}$Pu   (d) $^{2}_{1}$H

14. Which of the isotopes listed in Question 13 would you (   )
expect to have the lowest average mass per nuclear particle?

15. Which of the isotopes of Question 13 would produce the (   )
most energy during fusion with an identical nucleus?

16. In determining the age of organic material, you measure   ( )
  (a) the time required for half of the C-14 in the sample to decay
  (b) the ratio of C-14 to C-12 in the sample
  (c) the percentage of carbon in the sample
  (d) the time required for half the organic material to decay

17. In what order would you arrange the following reactions so   ( )
that the magnitude of ΔE per gram of reactant increased in that order?

$$\text{A.} \quad {}^{238}_{92}U \rightarrow {}^{4}_{2}He + {}^{234}_{90}Th$$
$$\text{B.} \quad {}^{236}_{92}U \rightarrow {}^{72}_{30}Zn + {}^{160}_{62}Sm + 4\,{}^{1}_{0}n$$
$$\text{C.} \quad {}^{2}_{1}H + {}^{2}_{1}H \rightarrow {}^{4}_{2}He$$
$$\text{D.} \quad H_2(g) + \tfrac{1}{2}O_2(g) \rightarrow H_2O(1)$$

  (a) $A < B < C < D$ (b) $D < C < B < A$
  (c) $D < A < B < C$ (d) $D < B < A < C$

18. The masses of ${}^{4}_{2}He$, ${}^{6}_{3}Li$, and ${}^{10}_{5}B$ are 4.0015, 6.0135, and   ( )
10.0102, respectively. The splitting of a boron-10 nucleus to helium-4 and lithium-6 would
  (a) evolve energy   (b) absorb energy
  (c) result in no energy change (d) cannot say

19. The decay of a neutron to a proton also yields a(n)   ( )
  (a) ${}^{0}_{-1}e$ (b) ${}^{0}_{+1}e$
  (c) α-particle   (d) ${}^{2}_{1}H^+$

20. The half-lives of radioactive ${}^{235}_{92}U$ and ${}^{238}_{92}U$ are 0.71×10^9 ( )
and 4.51×10^9 years respectively. For separate samples of these two isotopes containing equal numbers of atoms, the rate of decay is more rapid in
  (a) ${}^{235}U$ (b) ${}^{238}U$
  (c) both decay at the same   (d) impossible to say
      rate

Problems

1. For each of the following, write the symbol and give the charge and mass number.
  (a) beta particle   (b) positron
  (c) alpha particle   (d) gamma ray

2. Nuclear reactions occurring in stars are thought to include the fusion of helium nuclei to form such species as ${}^{8}_{4}Be$, ${}^{12}_{6}C$, and ${}^{16}_{8}O$.

   (a) Write a balanced equation for the fusion of $^4_2$He nuclei to give a $^{12}_6$C nucleus.

   (b) Calculate ΔE for the formation of a mole of $^{12}_6$C nuclei by this reaction (nuclear masses: ^{4}He = 4.001 50, ^{12}C = 11.996 71):
$$\Delta E(kJ) = 9.00 \times 10^{10} \, \Delta m(g).$$

3. Consider the decay of $^{210}_{84}$Po, which emits an alpha particle.
   (a) Write an equation for the reaction.
   (b) If the half-life for the decay is 140 d, what is the rate constant?
   (c) How long will it take for 90% of a sample of Po to decay?

4. A piece of wood found in a cave has a ^{14}C/^{12}C ratio of 5.75 × 10^{-13}. A piece of wood growing in the same area today has a ratio of 7.42 × 10^{-13}. If the half-life of ^{14}C is 5720 years, how old is the wood found in the cave?

5. The modern German and English languages are descendents of the Germanic language that split in two directions about 500 A.D. There is evidence that fundamental vocabularies change at such a rate that about 19% of the terms are replaced every thousand years. Estimate the percentage of the vocabularies that remain essentially unchanged in these two languages today. Describe an experiment *involving nuclear reactions* that would verify the date of 500 A.D. (more or less).

SELF-TEST ANSWERS

1. **F** (The reverse is true. Both mass number and charge must be conserved.)

2. **T** (Again, mass number and charge must balance.)

3. **F** (For example, many early workers in the field developed cancer.)

4. **F** (The rate, as measured by the rate constant, is inversely proportional to the half-life.)

5. **F** (The *change* in mass must be 1 g for this to be so.)

6. **F** (There may indeed be the risk of release of radioactive material, but an explosion like that in a nuclear bomb would require that the critical mass be brought together explosively.)

7. **T** (See the summary in Table 24.3 in the text.)

8. **d** (These are the charge and mass number, respectively, of the alpha particle.)

9. **b** (Some are very slow: consider the range of half-lives.)

10. **d** (The nucleus drops to a lower energy level.)
11. **d** (The total mass change is due to 7 alpha particles.)
12. **d** (Write out the balanced equation.)
13. **c** (To yield nuclei with a more stable mass per nuclear particle. See the discussion of Figure 24.3.)
14. **b** (Again, see Figure 24.3.)
15. **d** (Why?)
16. **b** (See, for example, Problem 4 below.)
17. **c** (Reaction A is a decay process. B is a fission; C, a fusion; D, an ordinary chemical change. You should be able to give orders of magnitude for ΔE in kilojoules per gram for each of these.)
18. **b** (The change in mass is positive.)
19. **a** (What is the balanced equation?)
20. **a** (See Question 4 above.)

Solutions to Problems

1. (a) $_{-1}^{0}e$ (b) $_{+1}^{0}e$
   (c) $_{2}^{4}He$ (d) γ (no mass, no charge)
2. (a) $3\,_{2}^{4}He \rightarrow {}_{6}^{12}C$
   (b) $\Delta m = 11.996\ 71 - 3(4.001\ 50) = -7.79 \times 10^{-3}$ g
   $\Delta E = 9.00 \times 10^{10} \times (-7.79 \times 10^{-3}) = -7.01 \times 10^{8}$ kJ
3. (a) $_{84}^{210}Po \rightarrow {}_{2}^{4}He + {}_{82}^{206}Pb$
   (b) $k = 0.693/140\ d = 4.95 \times 10^{-3}\,d^{-1}$
   (c) $Log_{10}(100/10) = 1.0 = (4.95 \times 10^{-3})(t)/2.30$
   $t = 460$ d
4. $k = 0.693/(5720) = 1.21 \times 10^{-4}$
   $Log_{10}(7.42/5.75) = 0.111 = (1.21 \times 10^{-4})(t)/2.30$
   $t = 2110$ years
5. First order reactions exhibit the property you are interested in: a given fractional change (-19%) occurs in a constant time interval (1000 years) (compare with a 50% change for $t_{1/2}$).

$$\log_{10}\frac{1.00}{1.00 - 0.19} = \frac{k(1000)}{2.30}\ ; k = 2.1 \times 10^{-4}$$

$$\log_{10}(X_0/X) = \frac{2.1 \times 10^{-4}\ (1500)}{2.30}\ , X/X_0 = 0.73$$

(In either language, 73% is unchanged.)

Experiment: use $_{6}^{14}C$ to date a manuscript in the parent tongue (which would have the same number of terms fundamental to both modern vocabularies). You need only find such a manuscript!

SELECTED READINGS

A more extensive treatment of nuclear chemistry:

Harvey, B.G., *Nuclear Chemistry*, Englewood Cliffs, N.J., Prentice-Hall, 1965.

Fission and fusion reactors, as well as their products and the interactions of these products with matter, are discussed in most issues of The Bulletin of the Atomic Scientists. *Also:*

Cromie, W.J., Which Is Riskier — Windmills or Reactors? *SciQuest* (March 1980), pp. 6–10.

Furth, H.P., Progress toward a Tokamak Fusion Reactor, *Scientific American* (August 1979), pp. 50–61.

Golub, R., Ultracold Neutrons, *Scientific American* (June 1979), pp. 134–154.

Hammond, R.P., Nuclear Power Risks, *American Scientist* (March-April 1974), pp. 155–160.

Kulcinski, G.L., Energy for the Long Run: Fission or Fusion? *American Scientist* (January-February 1979), pp. 78–89.

Lewis, H.W., The Safety of Fission Reactors, *Scientific American* (March 1980), pp. 53–65.

McIntyre, H.C., Natural-Uranium Heavy-Water Reactors, *Scientific American* (October 1975), pp. 17–27.

Olander, D.R., The Gas Centrifuge, *Scientific American* (August 1978), pp. 37–43.

Tandem Mirror, *SciQuest* (May-June 1979), pp. 24–26.

Vendryes, G.A., Superphenix: A Full-Scale Breeder Reactor, *Scientific American* (March 1977), pp. 26–35.

Walker, J., A Radiation Detector Made out of Aluminum Foil and a Tin Can, *Scientific American* (September 1979), pp. 234–246.

Weinberg, A.M., The Maturity and Future of Nuclear Energy, *American Scientist* (January-February 1976), pp. 16–21.

Nuclear reactions, as applied by a pioneer and misapplied(?) by one nation against another:

Bennett, C.L., Radiocarbon Dating with Accelerators, *American Scientist* (July-August 1979), pp. 450–457.

Cerny, J., Exotic Light Nuclei, *Scientific American* (June 1978), pp. 60–72.

Henahan, J.F., Glenn T. Seaborg, The Man from Ishpeming, *Chemistry* (December 1978), pp. 26–28.

Hersey, J., *Hiroshima,* New York, Knopf, 1946.

Impact: Interview with Glen T. Seaborg, *Journal of Chemical Education* (February 1975), pp. 70–75.

O'Nions, R.K., The Chemical Evolution of the Earth's Mantle, *Scientific American* (May 1980), pp. 120–133.

Seaborg, G.T., The New Elements, *American Scientist* (May-June 1980), pp. 279–289.

Wahl, W.H., Neutron Activation Analysis, *Scientific American* (April 1967), pp. 68–82.

On the synthesis of the elements:

Selbin, J., The Origin of the Chemical Elements, *Journal of Chemical Education* (May 1973), pp. 306–310.

25

AN INTRODUCTION TO
ORGANIC CHEMISTRY

QUESTIONS TO GUIDE YOUR STUDY

1. What properties are unique to organic compounds?
2. What elements are found in organic compounds? What kinds of interatomic and intermolecular forces are present in these substances?
3. How do you account for the vast number and variety of carbon compounds? How is carbon different from the other Group 4 elements?
4. How is it possible for two or more different compounds to have the same molecular formula?
5. How do isomers differ from one another? Are these differences explainable in terms of molecular structure?
6. How does the electronic structure of a bonded group of atoms, such as $\diagup$C=O, determine the physical and chemical properties characteristic of that group?
7. Having classified many organic compounds according to functional groups, what are the reactions characteristic of each class? Besides combustion, what other common reactions of organic materials can you think of?
8. What are the natural and synthetic sources of each class of organic compound? Are the natural sources renewable?
9. What synthetic products can you list which can be labelled as organic?
10. How do soaps and detergents work? What are their structural differences? What are their relative advantages and disadvantages?

YOU WILL NEED TO KNOW

Like Chapter 12, this chapter is mostly an extension of principles you have already seen in detail. A review of parts of Chapters 10 and 12 would be very useful, as suggested by the following tabulation.

Concepts

From Chapter 10, the following:

— How to write Lewis structures (see Skill 2 in the guide chapter).

— How to predict molecular geometry (see Skill 3).

— How to predict hybridization (sp, sp^2, and sp^3) (see Skill 5).

— How to indicate the number and kind of bonds (sigma and pi) (see Skill 6).

From Chapter 12:

— How to write molecular and structural formulas for alkanes, C_nH_{2n+2} (see Skill 2). This includes recognition of the general formula, writing all formulas for structural isomers.

— Draw resonance structures (see Skill 1).

— Compare physical properties of molecular substances (this is done for alkanes here).

Additional concepts include the following.

1. How to relate physical properties (bp, solubility) to intermolecular forces. See Skill 2 in Chapter 11.

2. How to draw structural formulas of geometric isomers. The concept is introduced in Chapter 19 (see Skill 2).

Math

1. How to write equations for acid-base reactions. See Skill 6 in Chapter 17.

2. How to represent weak acid and weak base equilibria; how to compare acid and base strengths. See Skill 4 in Chapter 17.

BASIC SKILLS

This chapter offers a change of pace. It illustrates several principles of molecular structure and reactivity in terms of selected organic compounds. You will see that most of the skills listed below are slight variations on skills you have already seen.

1. **Given the number of carbon atoms in a "straight"-chain alkane, alkene, or alkyne, give the molecular and condensed structural formulas.**

Example 25.1 illustrates this skill (like Skill 2 in Chapter 12) for an alkane. Example 25.2 does the same for an alkene and an alkyne. The exercise that follows Example 25.2 turns this skill around. Given the formula, you should be able to indicate the class of hydrocarbon being represented. The general formulas are:

alkane (all single bonds): C_nH_{2n+2}

alkene (one double bond): C_nH_{2n}

alkyne (one triple bond): C_nH_{2n-2}

2. Given the number of carbon atoms in a "straight"-chain alcohol or carboxylic acid, give the structural formula.

Like the preceding skill, the only new principle involved here is the recognition of a class of compound being represented. The alcohol functional group is $-OH$. The acid functional group is $-COOH$, or more explicitly, $-\overset{\text{O}}{\underset{\text{||}}{C}}-OH$. See Example 25.3. The exercise following it turns the skill around.

3. Given the formulas of an acid and an alcohol, write the formula of the ester they form.

The reaction in which an ester is formed is illustrated in the text by Equation 25.13. There it is pointed out that $-OH$ from the acid combines with $-H$ from the alcohol functional group to form water and the ester functional group. Another example:

$$CH_3COOH(aq) + CH_3CH_2OH(aq) \rightarrow CH_3COOCH_2CH_3(aq) + H_2O$$

Example 25.3 applies this skill.

4. Given the formula of a simple carbon compound, draw structural formulas for:
    a. all the structural isomers.
    b. all the geometric isomers.

The first part of the skill is illustrated by Example 25.4; the second part, by Example 25.5. Read again the comments in Skill 2 in Chapter 12. A systematic approach is needed in the application of this skill. Again, a molecular model kit would be very useful.

Geometric isomerism was first discussed in Chapter 19 in connection with the *cis-trans* isomers possible in square and octahedral complexes. Here, you see this kind of isomerism in connection with alkenes. Note that in order to have *cis-trans* isomers, the two groups attached to each side of the double-bonded carbon atoms must differ from one another. Thus the compound

$$\begin{array}{cc} H & H \\ | & | \\ Cl-C&=C-Cl \end{array}$$

can have a *cis* and a *trans* isomer

| *cis* | | *trans* |

but the compounds

$$\begin{array}{ccc} H\ H & Cl\ H & H\ Cl \\ |\ \ | & |\ \ | & |\ \ | \\ Cl-C=C-H, & Cl-C=C-H, \ \text{and} & Cl-C=C-Cl \end{array}$$

cannot show this type of isomerism.

Again, geometric isomers can exist for the compound $\begin{array}{c} H\ \ H \\ |\ \ \ | \\ Cl-C=C-Br \end{array}$ but not

for $\begin{array}{c} Br\ H \\ |\ \ \ | \\ Cl-\ C=C-H \end{array}$.

5. **Given, or having derived, the structural formula of a simple organic compound, identify each chiral carbon atom. Indicate whether or not optical isomerism is expected.**

Example 25.6 and the exercise that follows it apply this skill. Any time you find a chiral carbon atom in a formula, you expect the substance to exhibit optical activity. That is, there should exist two isomers that differ only in the sense of right and left hands. The criterion you apply in identifying a chiral carbon atom is that the atom be bonded to four different

atoms or groups of atoms, like so: $\begin{array}{c} d \\ | \\ a-C-b. \\ | \\ e \end{array}$

6. **Describe the properties of selected organic compounds.**

There is a considerable amount of descriptive organic chemistry in this chapter. In studying this material you should acquire a general knowledge of such topics as:

    — the various functional groups and some of their characteristic properties. (For example, −OH and −COOH, as well as NH, impart the relatively strong forces called hydrogen bonds. Some implications: high boiling points and high water solubilities.)

    — the processes used to obtain gasoline from petroleum.

    — methods used to prepare some common organic compounds, including alcohols and esters.

    — the characteristic structures and methods of preparation of soaps and detergents.

    A lot of this is summarized in the tables in this chapter.

SELF-TEST

True or False

    1. For each double bond in a hydrocarbon, subtract 2 H from ( ) the general formula for a saturated hydrocarbon, $C_n H_{2n+2}$.

    2. The molecular formula $C_5 H_{10}$ represents an alkene. ( )

    3. Among isomeric hydrocarbons, boiling point is expected to ( ) increase as the molecule becomes more compact.

    4. The hydrocarbon $C_4 H_{10}$ shows *cis-trans* isomerism. ( )

    5. Carbon always bonds to four other atoms. ( )

    6. The double bond in benzene and other aromatics behaves ( ) chemically in the same way as that in ethylene.

    7. Ethyl alcohol, $C_2 H_5 OH$, would be expected to show ( ) hydrogen bonding.

Multiple Choice

    8. Carbon compounds are so numerous and varied because ( )
        (a) C forms strong bonds with a variety of nonmetal atoms
        (b) C forms strong single as well as stronger multiple bonds with itself

(c) C atoms may bond in rings as well as in chains, branched and unbranched

(d) all of the above

9. The geometry of the four atoms in an alkyne $-X-C\equiv C-$ (   )
Y−, is

(a) linear   (b) bent
(c) square   (d) tetrahedral

10. Mere separation of petroleum into its components relies (   )
mostly on the process of

(a) fractional crystallization   (b) fractional distillation
(c) combustion   (d) cracking

11. The forces overcome in boiling acetic acid, CH_3COOH, (   )
include

(a) dispersion forces   (b) dipole forces
(c) hydrogen bonds   (d) all of these

12. The total number of isomeric alcohols with the formula (   )
C_4H_9OH is

(a) two   (b) three
(c) four   (d) more than four

13. Hydrogen bonding has an important effect on the physical (   )
properties of all but

(a) acids   (b) alcohols
(c) primary amines   (d) tertiary amines

14. Which one of the following is an ester?   (   )
(a) $C_2H_5-O-C_2H_5$ (b) $CH_3COC_2H_5$
(c) CH_3CH_2COOH (d) $C_3H_7COOC_2H_5$

15. Which one of the following should occur as *cis-trans* (   )
isomers?

(a) C_2H_4 (b) $C_2H_4Cl_2$
(c) $C_2H_2Cl_2$ (d) $C_6H_4Cl_2$

16. How many different compounds have the molecular (   )
formula C_5H_{12}?

(a) one   (b) two
(c) three   (d) some other number

17. The reaction $CH_2=CH_2 + HBr \rightarrow CH_3-CH_2Br$ would be (   )
called

(a) an acid-base reaction   (b) a saponification
(c) an addition reaction   (d) a substitution reaction

18. Large numbers and varieties of consumer products are (   )
derived from substances found in
 (a) air   (b) natural gas
 (c) graphite   (d) petroleum

19. The cracking of a mixture of alkanes may result in the (   )
formation of
 (a) H_2 (b) odd-electron molecules
 (c) higher molecular mass   (d) synthesis gas
      alkanes

20. Synthesis gas has potential value as raw material for the (   )
production of hydrocarbons primarily because it contains large
amounts of
 (a) CH_4 (b) CO_2
 (c) H_2 and CO   (d) synthetic gasoline

Problems

1. The molecular formula C_3H_8O may represent either an alcohol or
an ether. Draw a structural formula for each possible alcohol and give the
condensed structural formula alongside.

2. Draw structural formulas for all isomers, structural and geometric,
of C_4H_8. (Include no closed-ring structures.)

3. Designate each chiral carbon atom in the following formulas with
an asterisk(*).

$$\overset{\textstyle O}{\overset{\textstyle \|}{}}$$

 (a) CH_3CHCH_3 (b) $CH_3CHC{-}OH$
       $\mid$ $\mid$
       OH NH_2

4. Write a balanced equation for each of the following reactions:
 (a) acetic acid reacts with a strong base in water solution.
 (b) $CH_2{=}CH_2$ undergoes addition with H_2O at high T and P.
 (c) 1 mol benzene, C_6H_6(1), undergoes substitution with 1 mol
      Br_2(1).
 (d) ethyl acetate, $CH_3COOCH_2CH_3$, is saponified by heating with
      aqueous NaOH.

5. Both benzene and paradichlorobenzene, 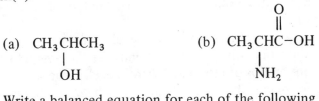 , are nonpolar.

Yet, hydroquinone, [structure of hydroquinone with OH and HO groups on benzene ring] , is polar. Explain these differences and

predict the relative melting points and water solubilities of these compounds.

SELF-TEST ANSWERS

1. **T** (For one double bond, as in an alkene, the general formula becomes C_nH_{2n}.)
2. **T** (See the preceding answer.)
3. **F** (The smaller, more compact molecule will have weaker dispersion forces. The bp will be lower. See Chapter 11.)
4. **F** (This is an alkane. There are none of the required C=C double bonds.)
5. **F** (Not, for example, when there are multiple bonds.)
6. **F** (Resonance hybrids "impart" or correspond to special bond properties of the benzene ring.)
7. **T** (H is bonded to the highly electronegative O. See Chapter 11.)
8. **d**
9. **a** (As predicted on the basis of electron pair repulsion.)
10. **b** (The nature of this process was first described in Chapter 1.)
11. **d** (Account for each of these.)
12. **c** (Can you draw all four?)
13. **d** (Where the functional group is $-\overset{|}{N}-$ and all sites attach to C atoms.)
14. **d** (Choice **a** is an ether; **b**, a ketone; **c**, an acid.)
15. **c** (The only one with a double bond and the possibility of two different groups of atoms attached to each carbon.)
16. **c** (The formula is that of an alkane. Start by drawing chains.)
17. **c** (See, for example, Equation 25.2.)
18. **d** (Especially in this age of petroleum-based plastics.)
19. **a** (As well as *lower* molecular mass alkanes, alkenes, and isomers.)
20. **c** (See, for example, Equation 25.7.)

Solutions to Problems

1. There are two isomeric alcohols. Showing the skeletons:

$$-\overset{|}{\underset{|}{C}}-\overset{|}{\underset{|}{C}}-\overset{|}{\underset{|}{C}}-OH \qquad -\overset{|}{\underset{|}{C}}-\overset{|}{\underset{|}{C}}-\overset{|}{\underset{|}{C}}-$$
$$\qquad\qquad\qquad\qquad\qquad OH$$

$$CH_3CH_2CH_2OH \qquad\qquad CH_3CH(OH)CH_3$$

2. There are four isomers, two of them geometric:

(cis)   (trans)

3. (a) none

(b) $CH_3 * CHC-OH$
         with $\overset{O}{\underset{\|}{C}}$ above and NH_2 below

4. (a) $CH_3COOH(aq) + OH^-(aq) \rightarrow CH_3COO^-(aq) + H_2O$
                                                    (Equation 25.12)
   (b) $CH_2=CH_2 + H_2O \rightarrow CH_3CH_2OH$ (Equation 25.10)
   (c) $C_6H_6(l) + Br_2(l) \rightarrow C_6H_5Br(l) + HBr(g)$ (Equation 25.5)
   (d) $CH_3COOCH_2CH_3(aq) + OH^-(aq) \rightarrow CH_3COO^-(aq) +$
         $CH_3CH_2OH(aq)$ (Equation 25.14)

   5. Consider the bonding geometry about the group C−O−H. Would you expect these three atoms to be coplanar with the benzene ring? Hydroquinone can exhibit hydrogen bonding and should be the most soluble in water as well as the highest melting.

SELECTED READINGS

Chemical bonding is considered in the readings of Chapter 10.

The history of organic chemistry is surveyed in:

Benfey, O.T., From Vital Force to Structural Formulas, Boston, Houghton Mifflin, 1964.

The scope of organic chemistry, from practical application to the origin of life:

Bailey, M.E., The Chemistry of Coal and Its Constituents, Journal of Chemical Education (July 1974), pp. 446–448.
Blumer, M., Polycyclic Aromatic Compounds in Nature, Scientific American (March 1976), pp. 35–45.
The Chemistry of Cleaning, Journal of Chemical Education (September 1979), pp. 610–611.
Clapp, L.B., The Chemistry of the OH Group, Englewood Cliffs, N.J., Prentice-Hall, 1967.
Kolb, D., Petroleum Chemistry, Journal of Chemical Education (July 1979), pp. 465–469.
Miller, S.L., The Origins of Life on Earth, Englewood Cliffs, N.J., Prentice-Hall, 1974.
Waddell, T.G., Legendary Chemical Aphrodisiacs, Journal of Chemical Education (May 1980), pp. 341–342.

26

ORGANIC POLYMERS; SYNTHETIC
───────────────── AND NATURAL

QUESTIONS TO GUIDE YOUR STUDY

1. What is a polymer? Name a few polymeric substances.

2. How do polymeric substances differ from the molecular and ionic substances studied so far? What are the differences in atomic-molecular structure? In physical and chemical properties of bulk samples?

3. What kinds of intermolecular forces are present in polymers? How do they determine physical properties?

4. What kinds of substances may react to form polymers? Are there correlations you can make with the Periodic Table?

5. What energy effects accompany polymer formation? What are the extents, rates, and mechanisms of such reactions? To what degree have we learned to control these processes on an industrial scale?

6. How would you experimentally determine the composition and structure of a polymer (e.g., the sequence of bonded atoms and their three-dimensional arrangement in a protein)?

7. What conditions (T, P, concentration . . .) prevail during polymerization reactions in biological systems? How do enzymes catalyze these reactions?

8. Do proteins and other polymers spontaneously form from simpler molecular units? Is life itself a spontaneous process?

YOU WILL NEED TO KNOW

Concepts

1. How to draw Lewis structures; predict molecular geometry; decide whether or not *cis-trans* isomers should exist. See Chapters 10 and 25.

2. How to predict the kinds of intermolecular forces and their effects on physical properties. See Chapter 11.

3. The nature of catalysis. A general knowledge is assumed here – see Chapter 14.

4. Some of the organic functional groups and their properties. See Chapter 25, particularly Table 25.4.

Math

1. The nature of weak acid, weak base equilibria. See Chapter 18.
2. How to work with pH. See, for example, Skill 2 in Chapter 17.
3. How to work problems in stoichiometry: relating formulas and molecular masses and % composition by mass. See Chapter 3.

BASIC SKILLS

This chapter is primarily descriptive. Few new concepts are introduced; no new calculations are performed.

1. For a polymer, relate the simplest molecular and structural formulas. Relate the molecular mass to the simplest formula and the number of monomeric units. Determine the percent composition by mass.

This is a complicated way of saying how some of the stoichiometric calculations of Chapter 3 are extended to polymers. Most of these calculations are straightforward, as seen in the several examples in this chapter. If you feel a little rusty in working them, look again at the related skills in Chapter 3.

Specifically, consider Examples 26.1, 26.5, and 26.6 as well as the exercises that follow them. Additional practice in this skill is given by some of the problems at the end of the chapter.

2. Given the formula of the monomer(s), write the structural formulas showing the expected dimer and part of the polymer resulting from
    a. addition.
    b. condensation.

An addition polymer is described by Example 26.2. Examples 26.3 and 26.4 each describe a condensation polymer and call for the formula of a dimer as well. Example 26.8 requires that you write one of two possible dimeric structures that result from a condensation reaction.

Note: As implied by the inclusion of Examples 26.3, 26.4, and 26.8 all under the heading of condensation, condensation polymers cover such diverse materials as polyesters, polyamides, polypeptides and proteins (interchangeable terms?), and many carbohydrates.

You should be able to work this skill in reverse. Given the structural formula for a dimer or a section of a polymer, give the formula for each monomeric unit.

3. Relate the formulas of an amino acid (in the forms of zwitterion, cation, and anion) to the relative pH of the solution it is in.

Again, this may sound more complicated than it really is. An amino acid contains both an acidic functional group, $-COOH$, and a basic functional group, $-NH_2$.

— At low pH, where there is excess H^+: the acid remains un-ionized, as $-COOH$; the weak base is converted to its conjugate acid, $-NH_3^+$. There is a net charge of 1+.

— At high pH, where there is excess OH^-: the acid loses its proton to give its conjugate base, $-COO^-$; the base remains as $-NH_2$. There is a net charge of 1−.

— At some intermediate pH, the predominant species is uncharged, but involves the transfer of a proton from the acid to the base, within the amino acid itself, to give the zwitterion (bearing the groups $-COO^-$ and $-NH_3^+$). It carries no net charge.

See Example 26.7 for an illustration of this skill.

4. Describe the preparation and properties of several dimers and polymers.

See, for example, the text discussion of polyethylene, polyvinyl chloride, rubber, nylon, sugars, and polysaccharides. Be able to interpret some of the properties in terms of three-dimensional structure and intermolecular forces such as hydrogen bonding.

SELF-TEST

True or False

1. Most synthetic polymers are solids at 25°C, 1 atm. ( )

2. The net photosynthetic reaction, $6\ CO_2(g) + 6\ H_2O(l) \rightarrow$ ( )
$C_6H_{12}O_6(s) + 6\ O_2(g)$, involves reduction of carbon.

3. The water solution of any amino acid which contains only ( )
one basic functional group and one acidic group is expected to have
a pH of 7.

4. The hydrolysis of a polyester is the reverse of a condensa- (   )
tion reaction.

5. The formation of a polypeptide from individual amino acid (   )
molecules is a condensation polymerization.

6. The α-amino acid $R-CH(NH_2)COOH$ is expected to exist as (   )
optical isomers.

Multiple Choice

7. The molecular mass of a polymer would probably be (   )
determined by
    (a) osmotic pressure
    (b) gas density
    (c) freezing point lowering
    (d) direct weighing of a single molecule

8. What products do you expect to form in the combustion of (   )
polyvinyl chloride?
    (a) CO_2 (b) H_2O
    (d) HCl   (d) all of these

9. The linkage between monomers in a polyester is: (   )
    (a) C−O−C   (b) C−O−C
                               ‖
                               O
    (c) C−N−C   (d) a hydrogen bond
        ‖ |
        O H

10. Addition polymers are expected to form from   (   )
    (a) F_3C-CF_3 (b) $F_2C=CF_2$
    (c) CF_4 (d) F_2

11. Condensation polymers are expected to form from   (   )
    (a) CH_3NH_2 (b) HCOOH
    (c) $H_2N-CH_2(CH_2)_4CO_2H$ (d) $HO-CH_2CH_2-OH$

12. Most synthetic polymers are ultimately derived from   (   )
    (a) wood   (b) natural gas
    (c) silicates   (d) petroleum

13. Many polymeric carbohydrates are condensation polymers (   )
of
    (a) glucose   (b) amino acids
    (c) ethylene   (d) phenol

14. The reaction of HO−R−OH and HO$_2$C−R'−CO$_2$H is ex- ( )
pected to give a
    (a) polyamide   (b) polyester
    (c) polypropylene   (d) polysaccharide

15. The helical structure of the protein in hair is a result of ( )
    (a) hydrogen bonding   (b) dispersion forces
    (c) genetic disease   (d) excessive shampooing

16. Which of the following carbohydrates cannot be utilized by ( )
the human body as a source of energy?
    (a) glucose   (b) sucrose
    (c) glycogen   (d) cellulose

17. Considering that the peptide linkage has the resonance ( )
structure
$$\begin{array}{ccc} C & & H \\ \diagdown & & \diagup \\ & C{=}N & \\ \diagup & & \diagdown \\ O & & C \end{array}$$
, the geometry of this group of atoms might be
expected to be
    (a) linear   (b) tetrahedral
    (c) planar   (d) octahedral

18. In a 0.10 M solution of glycine, H$_2$C(NH$_2$)CO$_2$H, at a pH ( )
of 10, the most abundant species (next to water) is
    (a) OH$^-$ (b) H$_2$C(NH$_3$)CO$_2$
    (c) H$_2$C(NH$_3$)CO$_2$H$^+$ (d) H$_2$C(NH$_2$)CO$_2^-$

19. The maximum number of different tripeptides that can be ( )
formed from three different amino acids, using one unit of each is
    (a) 2   (b) 3
    (c) 4   (d) 6

20. Fibrous structural materials found in mammals like you and ( )
me are likely to be composed of
    (a) carbohydrate   (b) cellulose
    (c) isoprene   (d) protein

Problems

1. Identify the monomers from which the following polymers were formed.

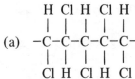

(a)

(b) $-\overset{\|}{\underset{O}{C}}-(CH_2)_2-\overset{\|}{\underset{O}{C}}-O-(CH_2)_2-O-\overset{\|}{\underset{O}{C}}-(CH_2)_2-\overset{\|}{\underset{O}{C}}-$

(c) $-\overset{H}{\underset{|}{N}}-(CH_2)_2-\overset{O}{\underset{|}{\overset{\|}{C}}}-N-(CH_2)_2-\overset{H}{\underset{|}{\overset{|}{N}}}-\overset{\|}{\underset{O}{C}}-N-$

2. Give the simplest formula for each of the polymers listed in Problem 1.

3. Vinyl alcohol, $CH_2=CHOH$, forms a polymer that is rather soluble in water.
    (a) Draw a portion of the polymer.
    (b) Explain the solubility, unusual among polymers.

4. Write a balanced equation for each of the following reactions.

    (a) Acrylonitrile, $CH_2=C\overset{\diagup H}{\diagdown CN}$, forms an addition polymer of n units.
    (b) Condensation of $HO-CH_2CH_2COOH$ gives a dimer.
    (c) The amide $CH_3NH-CO-CH_2CH_3$ undergoes hydrolysis.

5. The hydrolysis of sucrose, $C_{12}H_{22}O_{11}(aq) + H_2O \rightarrow 2\,C_6H_{12}O_6(aq)$, is catalyzed by $H^+(aq)$. In particular, the activation energy for the acid-catalyzed reaction is 107 kJ at 300 K. A certain enzyme found in yeast also catalyzes the hydrolysis, lowering the activation energy to 38 kJ. How many times faster is the enzyme-catalyzed reaction than the acid-catalyzed one? The rate constant and activation energy are related by the equation (Equation 14.13): $Log_{10}k = A - E_a/(19.1\,T)$, where A is a constant and E_a has units joules.

SELF-TEST ANSWERS

1. **T** (Reflecting strong intermolecular attractions between large molecules. See Chapter 11.)
2. **T** (What are the oxidation numbers? The reverse, a reaction involving O_2, is more obviously an oxidation.)
3. **F** (It depends on the relative strengths of the weak acid and the weak base.)
4. **T** (H_2O is inserted at each ester linkage, giving the alcohol and acid functional groups. Consider the reaction 26.5 in reverse.)

5. **T** (H_2O is eliminated for each peptide link formed. See Reaction 26.9.)

6. **T** (The alpha carbon atom is chiral. See Skill 5 in Chapter 25. What does this mean about the relative physiological activity of the different isomers?)

7. **a** (This would be the most sensitive method, giving the largest, most easily measured effect. See Chapter 16.)

8. **d** (The HCl makes it hazardous to burn.)

9. **b** (See the table of functional groups in Chapter 25.)

10. **b** (This would give the commercial product known as Teflon.)

11. **c** (The product would be like the nylons. Draw part of the structural formula.)

12. **d** (At least while the supplies last and are not too expensive.)

13. **a** (See Section 26.3.)

14. **b** (You need to recognize the alcohol and acid functional groups.)

15. **a** (See the text discussion of the pleated sheet and helical structures.)

16. **d** (We lack the necessary enzyme to catalyze the reactions.)

17. **c** (The geometry results from the double bond shown here. You should see that around either the C or the N, the geometry is triangular and planar.)

18. **d** (The acid should lose its H^+; the $-NH_3^+$ should also give up its proton. See Skill 3.)

19. **d** (Each amino acid could be at the $-NH_2$ end or the $-COOH$ end.)

20. **d**

Solutions to Problems

1. (a) $CHCl=CHCl$
   (b) $HO-(CH_2)_2-OH$ and $HOOC-(CH_2)_2-COOH$
   (c) $H_2N-(CH_2)_2-COOH$

2. (a) $CHCl$
   (b) Adding one of each monomer and subtracting H_2O per monomer gives $C_6H_8O_4$, which simplifies to $C_3H_4O_2$.
   (c) Each monomer added to a chain results in the loss of one H_2O: so, the simplest formula is $C_3H_7NO_2 - H_2O = C_3H_5NO$.

3. (a)
$$\begin{array}{cccc} H & H & H & H \\ | & | & | & | \\ -C-C & -C-C- \\ | & | & | & | \\ H & OH & H & OH \end{array}$$

(b) The $-OH$ group allows for hydrogen bonding.

4. (a)
$$n\left(\begin{array}{c} H \\ \diagdown \\ C=C \\ \diagup \quad \diagdown \\ H \quad\quad CN \end{array} \begin{array}{c} H \\ \diagup \\ \\ \end{array} \right) \rightarrow \left(\begin{array}{cc} H & H \\ | & | \\ -C-C- \\ | & | \\ H & CN \end{array} \right)_n \qquad \text{(See Equation 26.1)}$$

(b) $2\ HO-CH_2CH_2-COOH \rightarrow HO-CH_2CH_2-\overset{\displaystyle O}{\overset{\displaystyle \|}{C}}-O-CH_2CH_2-COOH +$

(See Equation 26.5)

(c) $CH_3NH-CO-CH_2CH_3 + H_2O \rightarrow CH_3NH_2 + HOOC-CH_2CH_3$

(See Equation 26.9)

5. For the enzyme-catalysis:

$$Log_{10}k_2 = A - \frac{38 \times 10^3}{19.1\ T}$$

for the acid-catalysis:

$$Log_{10}k_1 = A - \frac{107 \times 10^3}{19.1\ T}$$

Subtracting the second equation from the first and substituting 300 K for T:

$$Log_{10}(k_2/k_1) = \frac{(107-38) \times 10^3}{19.1(300)} = 12$$

$$k_2 = k_1 \times 10^{12}$$

The enzyme-catalysis is 10^{12} times faster!

SELECTED READINGS

Biological molecules, including DNA, and their chemical evolution, are discussed in almost any issue of Scientific American *and:*

Calvin, M., *Chemical Evolution,* New York, Oxford, 1969.

Calvin, M., Chemical Evolution, *American Scientist* (March-April 1975), pp. 169–177.

Dickerson, R.E., Chemical Evolution and the Origin of Life, *Scientific American* (September 1978), pp. 70–86.

Iversen, L.L., The Chemistry of the Brain, *Scientific American* (September 1979), pp. 134–149.

Watson, J.D., *Molecular Biology of the Gene,* Menlo Park, Calif., W.A. Benjamin, 1970.

Natural and synthetic polymers, their applications and preparations:

Billmeyer, F.W., *Synthetic Polymers: Building the Giant Molecule,* Garden City, N.Y., Anchor, 1972.

Chemistry (June 1978), the entire issue.

Connick, W.J., Flame Retardant Cotton Textiles, *Chemistry* (April 1978), pp. 13–16.

Cook, G.A., *Survey of Modern Industrial Chemistry,* Ann Arbor, Mich., Ann Arbor Science, 1975.

Epstein, A.J., Linear-Chain Conductors, *Scientific American* (October 1979), pp. 52–61.

Materials, Scientific American (September 1967).

Meloan, C.E., Fibers, Natural and Synthetic, *Chemistry* (April 1978), pp. 8–12.

Morton, M., Polymers – Ten Years Later, *Chemistry* (October 1974), pp. 11–14.

Uhlmann, D.R., The Microstructure of Polymeric Materials, *Scientific American* (December 1975), pp 96–106.

Witcoff, H., The Chemical Industry: What Is It? *Journal of Chemical Education* (April 1979), pp. 253–256.